William Bevan Lewis

The Human Brain

William Bevan Lewis

The Human Brain

ISBN/EAN: 9783337366292

Printed in Europe, USA, Canada, Australia, Japan

Cover: Foto ©berggeist007 / pixelio.de

More available books at **www.hansebooks.com**

THE

HUMAN BRAIN

HISTOLOGICAL AND COARSE METHODS OF RESEARCH

A MANUAL FOR STUDENTS AND ASYLUM
MEDICAL OFFICERS

By W. BEVAN LEWIS, L.R.C.P. (Lond.)

DEPUTY MEDICAL SUPERINTENDENT TO THE WEST RIDING LUNATIC ASYLUM

LONDON
J. AND A. CHURCHILL,
NEW BURLINGTON STREET

1882

PREFACE.

In these days of a voluminous scientific literature the appearance of every new manual demands some justification; a genuine want can alone excuse additions to an already extensive series of handbooks of scientific methods of research.

The *onus probandi* of this want rests, of course, with the author, and can in the present instance be readily met.

Prompted by the editors of *Brain*, as well as by a personally-felt need for a concise summary of methods of examination of the nervous centres, the author contributed to the pages of that journal a series of articles on "Methods of Preparing, Demonstrating, and Examining Cerebral Structure in Health and Disease."

An appeal thus made to what was regarded as a generally prevailing want amongst asylum medical officers and students, was met with signs of cordial approval; and the frequently expressed wish for the re-appearance of these articles in a separate and convenient form.

Such expressions of approval form the author's justification for this Manual of Methods.

West Riding Asylum, *April*, 1882.

CONTENTS.

PART II.

MINUTE EXAMINATION OF THE BRAIN.

                        PART III.

      LIST OF REAGENTS AND MOUNTING MEDIA.

LIST OF ILLUSTRATIONS.

——⋄——

THE HUMAN BRAIN.

HISTOLOGICAL AND COARSE METHODS OF RESEARCH.

INTRODUCTION.

IMPORTANT and necessary as are the minute investigations made by microscopic agency into the normal structure of the brain or its pathological deviations, we must guard ourselves from the very serious error of considering this method as essential or exclusively necessary in these studies. This error is a common one, and is too apparent to need much comment upon my part; for all who are engaged in the prosecution of cerebral histology must have recognized the prevailing tendency to disregard the naked-eye appearances of the brain, or to consider them as of very secondary import compared with a minute investigation aided by the complex armamentarium of the microscopist. If any of my readers are possessed of this notion I would ask them at the outset to rid themselves immediately of so fallacious an idea as one which, if fostered, must prove a serious obstacle to the acquirement of that intimate acquaintance with the true significance of the varied appearances presented by normal and diseased tissues. The skilled obstetrician recognizes as an invaluable acquirement that *tactus eruditus* which a constant and intelligent employment of a special sense can alone confer; and no less should the histologist endeavour to obtain a special visual tact, a highly refined and educated visual power, which can alone enable him to recognize by the unaided eye appearances which pass wholly unnoticed by the casual observer. I cannot too strongly insist upon this point, for he who would successfully study the morbid anatomy of the brain must, as in the morbid anatomy of other tissues, begin seriously to educate the eye to the coarse appearances presented to the unaided vision. The employment of a hand-lens of 2 to 4 inches focal length will prove of service here, the naked-eye appearance being contrasted with the magnified field, and the eye thus educated up to recognizing characters which without the aid of the lens were previously indefinite or unrecognizable. Nothing beyond repeated and energetic efforts in this direction will enlarge the area of the visual field and present to the mind the manifold appearances which constitute an unbroken whole, and which are absolutely necessary to a refined interpretation of the picture presented to the mind's eye. Repeatedly have I had occasion to observe that the student, after a full curriculum of hospital training— a training which should pre-eminently involve the high culture of the sense of sight, hearing, and touch—fails wholly to appreciate the most obvious abnormalities of the cerebro-spinal tissues, and this because he has neglected to tutor the eye so far as to learn what to look for, and how to look for it. If you place before him the brain of a case of chronic mania, without coarse lesion, and note his observations upon the appearances of a section across the hemisphere, it will be found that beyond a statement that the grey matter is pale, anæmic, and wasted, that the white matter is altered in consistence, and presents numerous coarse vessels, his untutored eye teaches him no more; and should he be examined even for the grounds upon which these statements are made, a too evident vagueness will be apparent—his ideas on relative depth of cortex or anæmic states of the brain and consistence of its tissue are gauged by no mental standard, and are remarkable only for their indefiniteness. The practised eye of the histologist, however, sees far more: the relative depth of cortex in various regions; the relation borne by this depth to normal standards; the varied depths of the several laminæ of the cortex, their distinctness of outline, general and local vascularity, as well as the deviations in hue dependent upon fatty or upon pigmentary changes. The œdematous, degenerated aspect of the medullary tracts, the presence of minute sclerosed patches, and a host of other morbid appearances, present to his mind a picture which the unpractised eye wholly fails to appreciate. To those who are liable to err in the direction pointed out, I now address myself in the hope of rendering their studies of the naked-eye appearance of healthy and diseased brain more inviting and instructive, by means of a few simple directions as to what appearances are to be sought and how they are to be looked for.

A few further statements on the object and scope of these articles may not prove amiss.

I address myself almost exclusively to the student who, up to the present time, has devoted but little of his attention to cerebral pathology or to the histology of the central nervous system; and more especially to asylum medical officers, whose opportunities for research in this field are so numerous, yet so sadly neglected. I shall endeavour here to place at the disposal of the student the more important, essential, and trustworthy information in this department which I find scattered promiscuously throughout an extensive literature. If amongst these gleanings my remarks should appear to the advanced student burdened by too much detail, my excuse must be that the primary object is to overcome the difficulties presented by these studies to the *novice*, difficulties which every expert observer has at the beginning to encounter.

PART I

COARSE EXAMINATION OF THE BRAIN

CHAPTER I.

General Anatomical Features.—It is necessary here that the student should recall to mind certain anatomical details affecting the relationships of this membranous investment of the brain which have an important pathological significance, and in the first place note that :—

1. The dura mater is a *fibro-serous* membrane ; the outer surface being fibrous and rough, the inner being smooth and polished by a layer of epithelial cells constituting a parietal arachnoid.

2. Cut off a small portion and float it in water. The rough pilose outer surface due to the numerous fibrous connections and vessels which unite it to the inner table of the skull becomes hereby very apparent in contrast to the smooth inner surface.

3. Next note that the dura mater is composed of *two distinct layers*, which, by their divergence, occasion the formation of the different venous sinuses ; whilst the inner layer, by its duplications, forms the various intra-cranial membranous partitions, the falx cerebri and cerebelli, as well as the tentorium.

4. Lastly, he must observe the close anatomical relationships between the lateral sinuses and the mastoid cells, and again between the superior petrosal sinuses and the internal ear. Caries of the petrous portion of the temporal bone and of its mastoid cells is so frequent an affection, that the contiguity to them of these venous sinuses is most important as

B 2

likely to occasion not only localized pachymeningitis or inflammation of the opposed dura mater, but inflammation and thrombosis of the venous sinuses, leading probably to metastatic deposits in the lung.

5. The position of the venæ Galeni, which receive the blood returning from the choroid plexuses and open into the straight sinus, is such that tumours or abscesses of the mid-lobe of the cerebellum would compress them.

Upon removal of the skull-cap the student should proceed to investigate the condition of the dura mater after the following systematic manner:—

1st. Note the general aspect of the membrane covering the hemispheres.

2nd. Examine the several venous sinuses.

3rd. Observe the condition of the dura mater in those regions at the base where there exists a special proclivity to disease.

General Appearance at the Vault.—A glance at the superficial aspect of the dura mater may reveal the presence of inflammatory products which are frequent here. If present, note their character, whether they are simple inflammatory exudates, capable of undergoing organization, or whether they be the results of suppurative inflammation. In the former case, observe how softened the membrane is, and how readily separable from the bone; in the latter case the presence of pus will almost certainly lead to the detection of a carious state of the internal table and diploë of the skull, and a more or less sloughy aspect of the disintegrating membrane. The organizable·blastema may be found in various stages of development, as loose areolar or thick tough fibrous tissue, or it may have formed bony plates which eventually unite with the cranial bones. Should dense fibrous bands be formed, adhesions betwixt the dura mater and bone occur, chiefly along the course of the sutures, or in localized islets over frontal or parietal bones, or so extensively that it is impossible to remove the skull-cap without at the same time including the dura mater, as their forcible separation would entail injury to the brain.

If there be the above given indications on the surface of recent or of old inflammation, note the *density*, *thickness*, *swollen appearance*, and *colour* of the membrane at these parts. Recent inflammations involve much swelling, interstitial infiltration, softening and looseness of texture, and a very red colour; whilst the effects of an old inflammation are seen in a callous fibroid induration and thickening, often with bony concretions formed in the interstices of its texture, together with the adhesion to the inner table of the skull just alluded to.

The student must be prepared to make allowances for the general increase in thickness and firmer adhesion of the dura mater, which pertains to advanced age independent of inflammatory changes. Superficial inspection will also enable him to detect attenuation of this membrane, such as results from the pressure of a hypertrophied brain, from the effusion of hydrocephalus or morbid growths. Local thinness of this membrane also may be due to the pressure of pacchionian bodies which perforate the dura mater frequently, and imbed themselves in fossæ on the internal surface of the cranium.

Extravasation of blood, forcing the membrane apart from the bone, may be frequently found, and its origin should be carefully sought.

Examination of the larger Venous Sinuses.—A fold of dura mater close to the longitudinal sinus at its exposed frontal extremity must now be pinched up by a forceps, and the scalpel passed through it and down into the longitudinal fissure of the brain, so as to divide the union betwixt the falx cerebri and crista galli. Introduce a curved bistoury into the anterior end of the longitudinal sinus, and carry the blade backwards to its occipital extremity so as to expose the sinus throughout its greater extent. The oblique orifices of the veins of the vertex which empty themselves into this sinus will now be exposed.

Note, 1st, The capacity of the sinus throughout.
Note, 2nd, The nature of its contents.
Note, 3rd, The condition of its lining membrane.

The incision through the frontal end of the falx cerebri should now be extended on either side through the dura mater backwards, upon a level with the sawn edge of the cranial bones. The anterior extremity of the falx being then seized with a forceps, should be drawn forcibly backwards from between the hemispheres, carefully dividing the great superficial veins where they open into the sinus, when the convolutionary aspect of the brain, covered by its soft membranes, will be exposed to view. At this stage of dissection the pia mater, arachnoid, and superficial aspect of the brain will require attention, and the method to be adopted will be detailed in the following chapter, to which the student must be referred. Continuing, however, our examination of the sinuses of the dura mater, it must be presumed that the brain has been removed, and the following procedure will then be necessary:—

Lay open the larger venous sinuses at the base, commencing at the torcular herophili behind, and carrying the blade down each lateral sinus to the foramen jugulare. Open up the straight sinus running into the torcular herophili in the base of the falx cerebri. Deal in like manner with the petrosal and cavernous sinuses.

A close inspection of these sinuses should be instituted with the object of learning the nature of their contents, and the morbid or healthy condition of the textures of these venous channels. The most frequent morbid conditions found here are those of phlebitis and thrombosis, the thrombus being usually the result of inflammation of the walls of the sinus.

Thrombosis of the venous sinuses of the dura mater is so frequent and so important an affection, that the possibility of its occurrence should always suggest itself when examining this membrane.

If a thrombus obstruct any one of these sinuses, note the more or less organized condition of the clot and its adhesion to the lining membrane of the sinus.

Examine closely its apparent site of origin and extensions, the latter often reaching into the jugular veins.

Next take into consideration the constitution of the clot, whether recent, organized and firm, or breaking up and presenting the appearance of a crumbling mass, which, during life, would give rise to metastatic deposits in distant organs.

The condition of the lining membrane and tissues of the sinus should next be noted. If inflamed, the inflammatory action should be traced to its origin, usually in a carious state of the cranial bones ; or by extension of simple inflammation of the dura mater in the neighbourhood of the sinus ; or less frequently, induced as a secondary result of thrombosis. The student should carefully study the effect of plugging of the different sinuses, and their proclivity to disease as a result of their varied anatomical relationships.

1. Inflammation of the *longitudinal sinus* will be accompanied by occlusion of the venous trunks from the convexity of the brain where they open into the sinus.

2. If the *straight sinus* be occluded as the result of adhesive phlebitis, the student will remark that its effects in obstructing the return of blood from the choroid plexus will be similar to what occurs when the venæ Galeni are compressed by a tumour of the mid-lobe of the cerebellum. Effusion into the ventricles might be expected in these cases, and the enlargement of the head from this cause has been recognized clinically.

3. If the *lateral sinus* be inflamed, we might expect, as a result, extension of phlebitis into the jugular veins, and plugging of the smaller sinuses or veins opening into it. In connection with this—note, that the small " emissary veins " (Santorini), which perforate the bone at the anterior extremity of the lateral sinus, and which form a communication *between*

the sinus and the *veins on the outside of the skull* at the back
of the head, may become plugged by a thrombus producing
a painful œdema behind the ear, also recognized clinically by
Griesinger.[1]

4. The position of the *petrosal* and *lateral sinuses* naturally
exposes them to phlebitis and thrombosis. Lying on the
petrous portion of the temporal bone, they are always liable
to be implicated by extension of inflammation from the
internal ear and caries of the temporal bone.

5. Thrombosis of the *cavernous sinus* has been frequently
noted. In connection with this condition the student must
be prepared to note the pressure which will probably result
to the carotid artery, which runs through this sinus invested
by its lining membrane.[2] He may also look for extension
of inflammation of the coats of this sinus to the facial veins
with which the ophthalmic vein communicates, noting the
obstruction to the return of blood from the cavity of the
orbit.

6. Amongst other *results* of thrombosis of the cerebral
veins and sinuses, will often be found punctiform hæmorrhages into the substance of the brain.

General Appearance at the Base.—Observe the condition of the dura mater at the base, where there exists a
special proclivity to disease. The sites alluded to are :—

1. Petrous portion of temporal bone.
2. Ethmoidal plate.
3. Parts adjoining superior cervical vertebræ.

---

[1] Quoted by Niemeyer in "Text-Book of Practical Medicine," vol. ii.
[2] This condition has been recognized clinically by a loud bruit heard on
auscultation of the skull. Dowse, etc. *Lancet*.

CHAPTER II.

General Anatomical Features.—In proceeding to investigate the healthy and morbid appearances of the arachnoid, the following anatomical facts will prove serviceable to
the student.

1. The arachnoid is usually regarded as consisting of a
parietal and visceral layer like ordinary serous sacs, the
former, however, being merely a layer of nucleated poly-
gonal epithelium lining the inner surface of the dura
mater; the latter forming the far more substantial investment of the convolutions of the brain separated from them
by the pia mater. This closed sac is lubricated by a
portion of the general cerebro-spinal fluid.

2. The visceral layer of arachnoid does not dip into the
sulci with the pia mater, but bridges them across, whilst in
other regions it is widely separated from the pia mater so as
to enclose between them extensive spaces. Thus, at the base,
a delicate veil of arachnoid stretches across the pons and
interpeduncular space as far as the optic chiasma; a similar
film of arachnoid closes in the fourth ventricle, stretching
across from the medulla to the cerebellum; the arachnoid also
does not pass down to the bottom of the longitudinal fissure,
but spans across immediately beneath the falx cerebri; hence
a space is left here between the two inner membranes of the
brain immediately above the corpus callosum.

3. The various spaces alluded to above all communicate
freely with one another, and with an extensive space around

the whole length of the spinal cord betwixt the pia mater and arachnoid sac. The whole system of spaces is termed the subarachnoid space, and is filled by the greater bulk of the cerebro-spinal fluid.

4. A space intervenes between the parietal layer of the arachnoid and the dura mater. It has been fully investigated and described by Axel Key and Retzius,[1] and is called by these anatomists the sub-dura-mater space (subduralraum). The student must therefore regard the membranous invest-ments of the brain as consisting of a firm and immovable fibrous outer membrane, the dura mater forming also the periosteal lining of the cranial cavity; of an intermediate serous sac or arachnoid whose visceral layer moves freely with the movements of the hemisphere; and lastly, of a vascular membrane supporting the blood-supply of the brain—the pia mater, which closely lines the cortex, dipping into the sulci, and being prolonged into the ventricles.

Between the various investments are four great cavities :—

1. The subduralraum betwixt dura mater and the parietal arachnoid.

2. The arachnoid cavity formed by the arachnoid sac.

3. The sub-arachnoid cavity betwixt visceral arachnoid and pia mater.

4. The epi-cerebral space betwixt pia mater and cortex.

It must also be borne in mind that compensatory adjust-ments for lessened or decreased intra-cranial pressure are obtained by means of the cerebro-spinal fluid; increase in the pressure, as in the increased amount of blood, growth of tumours, etc., being allowed for by escape of this fluid into the spinal sub-arachnoid space. In relation to this important subject it has been shown by Axel Key and Retzius, that the pressure of the cerebro-spinal fluid is always higher than that of the venous blood, and its specific gravity lower; hence its

---

[1] "Studien in der Anatomie des Nervensystems." Axel Key und Gustav Retzius. Stockholm, 1875.

free mingling with the blood-currents by osmosis is much facilitated.[1]

Coarse Examination of the Membranes.—The healthy and morbid appearance of the arachnoid should be subjected to close and critical examination.

Upon reflecting the dura mater the *contents* of the arachnoid sac will be revealed, and in relation to this point we must note :—

1. Variations in the amount of cerebro-spinal fluid here.

2. Modified appearance of this fluid from admixture with morbid products.

3. Note especially extravasations of blood, and the age of these hæmorrhages as revealed by the condition of the clot. It may be purely fluid blood, coagula, or a delicate film of fibrinous nature ; or it may be in various stages of organization. The various stages of the arachnoid cyst in relation to pachymeningitis have all been most graphically described.

4. Note the superficial aspect of the visceral and parietal arachnoid as regards—

   a. Absence of clear, moist, glistening surface.

   b. Presence of morbid deposits, such as films of lymph of varied appearance and stage of organization ; as a delicate, greyish, mucus-like exudation, or a membrane possessing a certain amount of consistence, or flakes which are yellow and puriform, and more rarely fibroid patches ; or, again, the development of bony plates in the parietal layer.

   c. Adhesions between layers of arachnoid sac, such as occur naturally by the development of the pacchionian bodies.

5. Note the anomalies of texture due to interstitial change and to deposits occurring here. The effect of frequent hyper-

[1] Op. cit.

æmia is seen in the opalescent, creamy white, or perfectly opaque appearance presented, especially along the course of the blood-vessels, and always seen after middle life to a greater or less extent. Hypertrophy of the textures, giving the membrane a tough and thick character, will result from similar causes. The membranes may be, however, soft, tumid, and swollen from œdema of its texture; and where it bridges the sulci a solid gelatiniform aspect is often given it by the subjacent fluid.

6. The relative amount of cerebro-spinal fluid in the sub-arachnoid space should next be taken into account. It should be collected, together with that which escapes from the arachnoid sac and ventricles, and carefully measured. The appearance of the arachnoid, when buoyed up by any undue amount of this fluid, should be noted before removal of the brain from the skull. It will be found that in these cases the arachnoid is widely separated from the pia mater by the subjacent fluid. The limpid character of this fluid may be tested by slitting up the arachnoid over a sulcus.

So far our remarks chiefly apply to the arachnoid, and we will now turn our attention to the pia mater. The proximity of these two membranes to one another of course predisposes to their common implication in any morbid process, yet the pia mater is not by any means unusually involved without extension to the brain-substance, and still more frequently without extension to the arachnoid. In healthy states these membranes are very thin and delicate, and removed from the brain-surface with difficulty. We should, therefore, note their *proximity* to each other over the summits of the gyri, alterations in the *thickness* of the pia mater and in its *toughness, infiltrations* of its texture, and the *ease or difficulty* experienced in its removal from the cortex.

Thus it may be found widely separated from the arachnoid, the latter being buoyed up by sub-arachnoid effusion. In these cases the appearance of the fluid should be noted, especially as regards the presence of inflammatory exudates, as flakes of lymph, etc.

If thickened in texture the alteration of the pia mater
should be traced to its origin, whether this be simple serous
infiltration or œdema conjoined with inflammatory material.
Tortuosity and varicosity of the vessels will also be further
indications of repeated congestions of this membrane. Especial caution must be observed against drawing hasty conclu-
sions regarding congestion of the membranes from the mere
fulness of the blood-vessels and hypostasis.[1]

The student will find ample opportunity for studying the
appearances presented by œdema and thickening of the soft
membranes of the brain in cases of senile atrophy, general
paralysis, and alcoholism.

The presence of inflammatory material modifying the
thickness and toughness of the membranes should lead to a
study of the *nature of the exudate*, whether it appears as a
tough exudate of plastic lymph more or less yellowish in
hue, or more serous, sero-purulent, opaque, and less organ-
izable material.

With the presence of these more or less plastic exudates
note the general infiltration of the membranes around with
greyish, opaque serum. Such changes afford excellent
opportunities for the study of the results of inflammatory
action in loose areolar tissue. In connection with inflammation of the pia mater—a condition which will frequently
engage the attention of the student—he should also follow up
his investigation of the *nature* of the exudate by examining
also :—

  a. The *direction* taken by the morbid product, noting
       that this occurs almost invariably along the vas-
       cular tracts, the course of the vessels being marked
       out by opacities.
  b. The *extension* to neighbouring structures (brain,
       arachnoid, dura mater, and skull).
  c. The *limitation* of inflammation to *small areas* (as
       when originating secondarily from caries of the

---

[1] On this point see Niemeyer. Art. "Hyperæmia of Brain and its Membranes." "Text Book of Medicine," vol. ii.

cranial bones) ; or to the *convexity* of the hemispheres (chiefly accompanied by plastic products) ;
or, lastly, characterized by the *base* of the brain
being originally and chiefly involved (usually
accompanied by aplastic and tubercular products.)

Note attentively the greater or less facility of stripping
the pia mater from the cortex. The membrane may be
œdematous, thickened, gelatinous, and most readily removed ;
or, on the other hand, it may cling with greater tenacity or
adhere so firmly that portions of the cortex tear away with it,
leaving an eroded, worm-eaten aspect of the surface of the
gyri.

Attention should be directed to the strength of these adhesions, their implication of the summits of the gyri only, their
localization over the convolutionary surface of the brain, and
the coarseness of the blood-vessels entering the cortex at
the sites of adhesion.

Whenever the student is engaged with a case of meningitis, the morbid topography must be carefully studied. If
basic, the veil of arachnoid extending from the optic chiasm
to the pons must be examined and removed, noting implication of the numerous vascular and fibrous twigs extending
betwixt it and the subjacent pia : follow these results of
inflammatory action up along the vessels of the fissures of
Sylvius and the anterior cerebrals in the longitudinal fissure.
Examine the regions just named for tubercular granulations,
paying especial attention to the smaller blood-vessels, which
may have tubercular masses in their sheaths, occluding more
or less the calibre of the vessel. Should granulations appear
around the blood-vessels, let them be removed and floated in
water for more careful observation.

CHAPTER III.

Dissection.—After carefully noting the external appear-ance of the membranes covering the hemispheres both at the vertex and the base, the condition of the arterial system should be inquired into. The brain being placed with its base uppermost, the fine expanse of arachnoid which bridges across the inter-peduncular space is first removed, and the great vessels forming the circle of Willis will then be ex-posed to view. Now separate the temporo-sphenoidal lobe from the adjacent frontal and parietal lobes by dividing the bridge of arachnoid extending between them across the Sylvian fissure.

On gently drawing back the temporo-sphenoidal lobe the Sylvian branch of the middle cerebral artery will be observed running deeply in the fissure towards the upper extremity of the latter, giving off small branches in this course. The radiating gyri of the island of Reil (central lobe) will in most cases be only partially exposed, and it will be necessary to separate the operculum, or that portion of the third or inferior frontal convolution which laps over the anterior portion of this lobe. The five or seven gyri of the island will now be seen radiating outwards like a fan and supporting the various branches derived from the middle cerebral in its sulci. Split up the arachnoid which bridges across the longi-tudinal fissure anteriorly from one frontal lobe to the other. We shall thus have exposed the various branches of the circle of Willis as far as they can be seen at the base. Note first the *arrangement of the blood-vessels.*

The Circle of Willis.—A first glance at the arrangement of the great arteries at the base of the brain entering
into the formation of the circle of Willis cannot fail to
impress upon the student's mind this important fact: there
are here *two great arterial systems more or less distinct and independent*. In front lies the *carotid system*, supplying by far
the greater bulk of the brain ; behind lies the *vertebral system*,
distributed to the posterior and inferior regions of the cerebrum—an area small in comparison to the former. These
two great arterial systems are united by the two posterior
communicating arteries which connect each carotid with its
corresponding posterior cerebral artery.

Examine the posterior communicating arteries and note
the remarkable smallness of their calibre, a fact which suffices
to ensure us that the circulation in the carotid and vertebral
systems *is in the main distinct ;* whilst the arrangement
allows of compensatory enlargement and a free communication betwixt both systems where emboli, thrombi, or diseased
textures obstruct the circulation of a main branch.

Next observe the junction between the two anterior cerebrals deep in the longitudinal fissure by means of the cross
branch, the anterior communicating artery. Raise the branch
on the forceps and note its short length, small calibre, and
right-angled direction, facts which teach us that, although in
case of obstruction on the carotid side of the anterior cerebral,
this branch, by dilating, may afford a satisfactory re-establish-
ment of circulation over the area to which it is distributed,
yet for the greater part the circulation *betwixt the two anterior
cerebrals is distinct*.

A little further consideration will teach us that the circulation of *each hemisphere* is by the above mechanism rendered
wholly *distinct and independent ;* that the circulation of the
mid-cerebral region of one side is *wholly separated* from that
of the opposite hemisphere ; whilst the anterior cerebral areas
are *more closely associated* through the medium of an anterior
communicating branch.

Lastly, through the medium of the posterior communicating artery (especially when we recall to mind the frequent enlargement of this vessel met with) there will be a far more ready communication betwixt the carotid and vertebral circulation than there can be betwixt the vascular apparatus of both hemispheres. We should pay especial attention, therefore, to the following observations:—

a. The almost complete independence of hemispheric circulation.

b. The very complete independence established betwixt the circulation of one middle cerebral and that of the other.

c. The interdependence established betwixt the two anterior cerebral streams through the medium of an enlarged communicating branch.

d. The possible admixture of the carotid and vertebral circulation of the same side through the medium of the posterior communicating—*a condition very frequently established.*

The arrangement of the internal carotid ere it reaches the circle of Willis is one of interest and significance. Within its bony canal the tortuous sigmoid course taken by it is undoubtedly one means whereby the brain is protected from the results of the cardiac pulsations. The student will recall a similar tortuous course taken by the vertebral artery ere it enters the cranium. It has long been noticed that the middle cerebral is more readily blocked by emboli than the other branches, and that the area of its distribution shows especial proclivity to hæmorrhage. Anatomical reasons, and the fact that this vessel lies "more directly in the way of strain from the heart explains its frequent plugging and rupture from disease" (Hughlings Jackson[1]). Thus Prevost and Cotard found that tobacco seeds injected into a dog's carotid most often lodged in the middle cerebral. Again, the fact that the left carotid arises directly from the summit of the aortic arch, whilst the right, arising from the innominate,

[1] *Lancet*, September 4, 1875.

is inclined at an angle to the aortic current, is sufficient to
explain the greater immunity from embolism experienced by
the right carotid in its distribution to the brain. The right
vertebral arises from the horizontal part of the subclavian,
and is therefore also less subject to embolism than the left
vertebral, which arises from the summit of the ascending
portion of the left subclavian.

The Arterial Tunics.—Note the following points :—

    1. Colour and opacity.
    2. Relative toughness of coats.
    3. Tortuosity or kinks in the course of the vessel.
    4. Local bulgings or constrictions from diseases of texture.

1. The arteries may be perfectly white in cases of slow
and lingering death, and contain a decolourized clot extending into its minute ramifications. On the other hand, they
may be of a deep red hue from fluid blood, dark clots, and
blood staining of the lining membrane. The vessels, again,
may be thin and semi-transparent, or opaque from thickening of the arterial tunics and morbid deposits.

2. The toughness of the arterial tunics can be tested later
on after their removal. Their resistance to strain is an important feature, and due attention should be paid to it by the
student. The apparent thinness of a vessel should never
be allowed to deceive him, as the thicker vessels are often
most degenerated and least resistant to traction or expansion.
Traction may be applied to the vessel held between two pairs
of forceps, when an approximate idea of the breaking strain
may be acquired, whilst the use of a conical gauge will
enable him to note the various degrees of resistance these
textures offer to expanding forces.

3. Tortuosity and kinks in the smaller branches should be
noted as suggestive of previous forcible distension from congestion. This is often apparent in the larger arteries ; thus,
Quain has noticed frequent tortuosity of the internal carotid
before it enters the carotid canal outside the skull in apoplectic subjects.

4. Local bulgings may be due to aneurismal dilatations, to the different forms of arteritis—atheroma, tubercle, syphilitic gummata, or to the impaction of a thrombus or embolus. The arteries at the base will be found often exceedingly atheromatous, tortuous, knotty, and white ; the amount of atheromatous material occluding even the larger cerebral arteries, often converting the smaller branches into irregular knotty cords. The student should never infer from the aspect or feel of such a vessel that it is occluded, but he should always make a section across the mass of diseased tissue, when, if the vessel be permeable, the orifice is readily detected.

Note particularly the branches which are occluded. If the bulging be due to inflammatory swelling and exudation into the outer tunics of the artery, it will be observed that corresponding to the site of lesion the vessel is *dilated*—a condition due to paralysis of the muscular coat of the vessel, together with implication of the elastic outer tunic. Make a section across the inflamed tissue, and observe how readily the elastic or outer coat may be stripped from the muscular by means of a forceps; also how friable and easily separable are the muscular fibres of the media. The inner coat will be probably deeply stained by hæmatin, or may be eroded and covered by an adherent clot immediately beneath the inflamed patch. When a clot appears to obstruct the calibre of a blood-vessel, the character of the clot should be taken into consideration, its fibrinous constitution, stage of organization, adhesion to vascular walls, its form, prolongations, and appearance of its section. The sheath and outer tunic of the arteries should be closely examined, especially in the neighbourhood of the Sylvian fissure and island of Reil, for tubercular or syphilitic growths. Especial care must be taken to exactly note the arterial tunics involved, so as to discriminate between ordinary *endarteritis*, resulting in atheromatous degeneration, and the syphilitic *node* or gumma involving the outer coats, and the tubercular nodules involving the arterial sheaths, all of which lesions may give rise to blocking, con-

traction of the cavity, and thrombosis. Such lesions should
be examined microscopically. In all cases the above superficial examination should be supplemented by removal of
short lengths of any diseased vessels, to be reserved for
freezing and section-cutting, according to the methods
detailed in the microscopic section.

The vessels spreading over the island of Reil and up the
Sylvian fissure should now be raised by passing a forceps
beneath them, and with a clean sweep of the scalpel they
should be all divided just where they turn over on to the
superficial aspect of the brain ; this should be repeated on the
opposite side. The two anterior cerebral arteries should now
be divided at the genu of the corpus callosum, and dissected
back to their origin from the carotids. The posterior cerebrals will be found running backwards round the crura
cerebri under cover of a bridge of arachnoid. The posterior
cerebrals, together with the three cerebellar branches, may
be followed to a short distance and then divided. It will
now be found that the circle of Willis is retained merely by
the numerous minute nutritive branches, between the basilar
artery and pons, and those from the anterior, middle, and
posterior cerebrals, which pass through the anterior and posterior perforated spaces. Gently draw these vessels out by
means of a forceps as far as possible, sever them close to the
surface of the brain, and then float off the vessels of the base
into a shallow dish or plate of water, arranging the branches
in their relative positions.

Capacity of the Arteries at the Base.—It may be
found advisable in certain cases to test the relative capacity
of the larger vessels at the base, and this may be obtained
approximately by means of a small graduated conical gauge.

In making these comparisons the student should bear in
mind that the *united areas of the branches* equal very nearly
the *area of the trunk* from which they originated, although
their united diameters far exceed that of the latter. It has
been shown by Paget's measurements that the equality

between the area of the trunk and of its branches is not *exactly* maintained, the area sometimes increasing in the vessels of the upper extremities and head and neck, and diminishing in the lower extremities. Now, according to the mathematical law that the areas of circles are as the squares of their diameters, it will be necessary to contrast in our measurements the square of the diameter of the trunk with the sum of the squares of that of the branches arising from it. The student, therefore, should proceed in the following manner :—

Let us suppose he wishes to estimate the relative capacity of the vessels at the base and their primary branches ; let him remove these vessels as already recommended, and float them out in a shallow vessel containing water. With a sharp pair of scissors cut across the vessel exactly at right angles to its direction, pass the graduated cone into it, and gently draw the vessel on until it is fully *distended*, but not *stretched*, by the gauge ; read off the diameter as shown on the gauge, and proceed in like manner with the other vessels. Arrange the diameter and the squares in appropriate columns, opposite the names of the vessels, for future comparison. The following tables represent the average measurements of the various vessels at the base in forty-five cases of insanity I have investigated :—

	DIAMETER.	SQUARE.
	mm.	mm.
Vertebral Artery—Right . . .	3·147	10·341
,, ,, Left . . .	3·42	12·039
Basilar	3·82	14·805
Posterior Cerebral—Right . . .	2·658	7·213
,, ,, Left . . .	2·56	6·551
Carotid—Right	3·951	15·789
,, Left	4·02	16·026
Mid-Cerebral—Right	3·133	10·063
,, ,, Left	3·55	10·851
Anterior Cerebral—Right . . .	2·73	7·55
,, ,, Left . . .	2·66	7·445

Thus the sum of the square of the diameters of the anterior and middle cerebrals of the right side was to that of the left side as 17·613 to 17·796.

The square of the diameter of *right and left* anterior and middle cerebrals amounted in the aggregate to 35·409, both posterior cerebrals reaching only 13·764.

Contrasting again the mid-cerebral supply of both sides with that of both anterior cerebrals, we find it represented by 20·414 against 14·995.

Reference to the table of measurements will show that the posterior and anterior cerebral supply of both sides is almost exactly similar: thus, on the right side, we find 7·213 contrasted with 7·55; and on the left side, 6·551 with 7·445.

Another interesting feature in connection with these measurements is the great preponderance in areas of the two vertebrals over the basilar, which they form by their union. The two former have as the sum of the square of their diameters 22·38, the basilar only averaging 14·805; and in several cases I have found the basilar artery not more than one-half the area of the united vertebrals.

It may assist the student to obtain a clearer idea of these areas if he represents them graphically as straight lines upon an enlarged scale.

Vessels of the Pia Mater.—First note the general appearance and distribution of the vessels of the pia mater as they lie *in situ*. Observe the relative position taken by the veins and arteries, the former being large and superficial, the latter much smaller and concealed chiefly within the folds of pia mater, which dip into the sulci and are bridged across by arachnoid. The upper aspect being arachnoid, should, in health, present a smooth endothelial surface. *Use a hand-lens in examining these vessels.* The student should observe any fibrinous effusions, purulent exudates, or minute extravasations along the course of the blood-vessels, the presence of atheroma, morbid growths—as gummata, tubercle, etc. If tuberculosis be suspected, examine the arteries deep in the sulci, and this with care, as tubercle along the course of the vessels is found with far greater difficulty here than at the base or up the Sylvian fissure.

Next strip off a large portion of arachnoid and pia mater from the surface, and transfer it to a deep vessel containing water, the arachnoid surface being uppermost. Observe the velvety aspect of the lower surface, due to a fine pile of blood-vessels, and the very rich fringe of vessels along the inter-gyral folds, for the supply of the cortex deep within the sulci.

Seize with a forceps one of these richly fringed folds and snip off a portion with a scissors, floating it on to a glass slide, with its vessels disentangled and spread out. Remove also to another glass slide a portion of the membranes detached from the summit of a convolution, and study them by means of a hand-lens, both in reflected and transmitted light. Educate the eye in this manner to the recognition of the various healthy and morbid structures to be found here.

For microscopic purposes, both these slides last mentioned should be reserved, whilst a small portion of brain covered by its membrane, and including two *adjacent* convolutions, should also be set aside. These will afford subjects for a *bird's-eye view* of the surface under low powers of the microscope, such an examination being always insisted upon as especially valuable. The portion of brain must then be frozen and cut in such a direction that the sections so obtained represent both gyri with the intervening sulcus bridged over by arachnoid. We thus obtain very beautiful slides for fresh examination, which afford us the opportunity of minute examination of the cortex and the connections and relationships between it and its membranes.

The large superficial veins at the vertex where they open into the longitudinal sinus, are occasionally found plugged by thrombi; they should therefore always be examined with the object of ascertaining the nature of their contents, especially if there is evidence of inflammation of the dura mater or its sinuses, or of purulent meningitis.

The student should also take every opportunity of following up the arterial branches given off from the anterior, middle, and posterior cerebrals, in their course along the sulci, until he has familiarized himself with the exact supply of various regions and convolutions of the brain. It is only necessary to allude to the work of Duret and Charcot in this direction to indicate the importance of a minute knowledge of the arterial districts of the encephalon in order to fully appreciate numerous lesions met with in the brain.

Nutrient Vessels at the Base.—We should devote attention to the numerous vascular tufts which arise from the large vessels at the base for the supply of the central ganglia and pons Varolii. Those which supply the basal ganglia are disposed into anterior and posterior groups—the former arising from the anterior cerebral, anterior communicating, and middle cerebral ; the latter taking origin from the bifurcation of the basilar, often from an enlarged posterior communicating, a further series arising from each posterior cerebral artery as it winds around the crura cerebri. Roughly, it may be stated that the anterior group supplies the corpus striatum and anterior extremity of the thalamus opticus, whilst the posterior group is distributed chiefly to the thalamus.

Note, first, their remarkably *fine calibre*, considering their direct origin from main arterial trunks ; and secondly, observe that they are given off from these trunks at a *right angle*—a significant fact, as they can only be affected to a minimum degree by the propulsive wave of blood caused by each cardiac systole. They are filled, therefore, by a sustained lateral pressure.

An examination of the nutrient supply to the pons will impress us with the following facts :—

    a. The nutrient twigs arising from the main artery are remarkably small, as in the other cases.

    b. They also arise at right angles from the basilar.

    c. The basilar is but slightly over half the capacity of both vertebrals, which supply it.

The latter fact, which I have already demonstrated by measurement in our remarks upon arterial capacity, when taken in conjunction with the other two data, warrants us in assuming that these nutrient branches are kept constantly filled by high lateral pressure, and exhibit none of the phenomena of the pulse. I need not insist upon the beauty of this arrangement of means to end.

In removing the vessels at the base, be careful not to tear these nutrient vessels short, but with gentle traction draw them out and cut them across with fine curved scissors ; float the circle of Willis and its branches in a shallow dish of water, and examine the various nutrient groups. Snip off some of the anterior and posterior tufts, float them on to a slide, arranging the branches, and examine by the naked eye as well as by microscopic aid.

Wherever arterial degeneration is suspected, never miss the opportunity of thus closely inspecting the nutrient supply of the corpora striata, optic thalami, and pons Varolii. The naked eye will suffice to reveal to us aneurismal dilatations along these branches, atheromatous degeneration, thrombi or embolic plugging, often explanatory of hæmorrhages within the structure of the basal ganglia or pons.

CHAPTER IV.

PHYSICAL PROPERTIES OF GREY AND WHITE MATTER.

In the introductory chapter to his great work on Pathological Anatomy, Rokitansky has said :—"Just as there is a general and a special anatomy, physiology, pathology, so there must in like manner be a general and a special pathological anatomy. The former treats of anomalies of organization, the latter of the special anomalies of individual textures and organs.[1] " It appears to me that in the examination of the diseased brain this natural classification must be kept in view. The student should be taught to appreciate the various general anomalies of organization as they present themselves in the brain; and the methods to be adopted in estimating the degree, extent, and significance of such physical deviations. It is, of course, not our function to detail the various diseases of texture to which the brain is subject, but rather to place the student in possession of those indications of healthy and diseased tissues which are presented by alterations in the physical properties of grey and white matter. The elements of a physical diagnosis, so to speak, are placed at his disposal, and hints thrown out for his guidance, where it is thought a wrong construction might be attached to the physical signs.

It will be found convenient, however, to illustrate the more profound anomalies in consistence, colour, specific weight, and other physical qualities by reference to some of the *special diseases* of *texture* to which the nervous centres are liable, and

[1] " Pathological Anatomy," Sydenham Soc. vol. i.

which may serve as typical cases wherein the student may acquire a practical acquaintance with the methods of physical research.

The fundamental physical properties of the nervous tissues with which it is necessary we should acquaint ourselves are briefly as follows:—Consistence, colour, volume, and weight, both absolute and specific. Each in turn will now claim our close attention.

§ CONSISTENCE.

The firmness, solidity, or consistence of a texture depends upon the cohesive force exerted by the individual elements of its mass, and to which is due, on the one hand, the varying degrees of toughness, hardness, and resistance, or, on the other hand, of softening and friability and liquefaction, to which they are liable. The consistence of a compound tissue so complex as the brain is necessarily subject to very great variations, since its individual constituents vary amongst themselves greatly in their physical properties. An organ composed of nerve-tubuli, nerve-cells, of a connective framework, and an elaborate system of arteries and veins, the whole pervaded by nutrient fluid, must be subjected to great modifications in consistence due to alterations in the relative proportions of one or other of the constituents. Hence it is that we get such variety in the structural firmness of brain-tissue, not only in different animals and in the brain of man, but in different regions of the same brain. It is this textural difference betwixt grey and white matter which explains their varying degrees of consistence, the same explanation pertains to the firmness of the occipital as contrasted with the frontal lobe, and of the central grey ganglia as contrasted with the grey envelope or cortex of the cerebrum. The grey cortex, consisting of a vast assemblage of nerve-cells and their protoplasmic extensions, imbedded in a most delicate web of connective tissue, with a complicated vascular apparatus, has a far lower degree of consistence than the white matter which owes its solidity to large medullated nerve-fibres and an abundant and coarser connective matrix.

A moment's consideration will suffice to indicate that the forces which in disease modify textural cohesion must come from *without* or from *within* the structural elements; in other words, the tissue elements may be forced asunder by fluids pervading the texture and by the intrusion of adventitious bodies, or by nutritive changes in the elements themselves, whereby a quantitative or qualitative transformation may be induced.

Conditions modifying Normal Consistence.—1. The forcible infiltration of nervous texture by serous fluid, by plastic exudates, or by blood, will in the first place tend to destroy all textural cohesion, and consequently produce reductions in consistence; whilst further changes in the effused blastemata due to organization will result in increased firmness or induration.

2. Chemical changes in the constitution of the individual elements, according to their nature, may tend to produce increase or reduction of consistence, *e.g.*, fatty, amyloid, and calcareous degenerations, the nutritive anomalies due to inflammation, decomposition of the structural elements due to putrefactive changes.

3. Apart from the above qualitative anomalies of nutrition, we must not overlook the quantitative element which sometimes exists alone or in combination with the former.

Such are the forms of hypertrophy and atrophy due to an *increase* in *size* and *number* of constituent elements on the one side, and on the other, in the *decrease* in *size*, *wasting*, and *diminution in the number* of the tissue elements.

The student must guard against the fallacy of regarding density and consistence of a tissue as in any way mutually convertible terms, for although in a large number of cases increased density may co-exist with increased consistence of the brain (sclerosis, hypertrophy, etc.), yet conditions occur where with increase of density there is really a diminution in consistence. Thus, in inflammatory conditions of the brain-tissue where plastic exudates have forcibly reduced the cohesion of the mass, the *specific weight* of the part is notably increased. Hence the specific gravity of cerebral tissue will afford us no exact and reliable guide as to degrees of consistence; and, in fact, we have no more exact gauge of consistence of texture than the rough-and-ready methods afforded by the sense of sight and touch.

Estimation of Textural Cohesion.—1. By the eye we note deviations from the natural compactness of an organ, or the maintenance of its normal contour. Contrast the brain of a senile dement or of a general paralytic with the compact firm brain of epilepsy; or the same organ in warm weather and in a state of incipient decomposition with the recent brain removed from the skull in cold frosty weather.

2. By the sense of touch note the relative resistance to pressure and the compressible and lacerable character of the tissues. Especially note this with respect to the central commissural tracts, basal ganglia, fornix, and septum lucidum.

3. By its resistance to a graduated force, such as the rough gauge afforded by a stream of water falling from a variable height (applicable only to cases of much reduced consistence). Any case of white or inflammatory softening may thus be tested.

4. By its resistance to section (applicable to the slighter degrees of reduction and all degrees of increased consistence). Contrast the resistance to section presented by the medulla, pons, cerebellum, cerebral hemispheres, and ganglia at base.

Observe that we gain by these means not only a satisfactory gauge of consistence, but that the methods vary in their relative value in different cases. We should take every opportunity of rendering ourselves perfectly familiar with the general consistence of healthy brain prior to post-mortem change. Let us remove a normal brain for careful study, and note the following points:—

The Normal Brain.—Observe that upon removal from
the cranial cavity it preserves its normal contour, although it
has lost its natural support—the cranial bones. It is plump,
rounded, and compact, giving a general impression of firmness
and solidity, ere we gauge its consistence further by handling.
The hemispheres are closely approximated by means of the
strong commissural band—the corpus callosum, showing no
tendency to fall apart from rupture of the commissure, such
as is so frequently seen in the degenerated and diseased brain.
Upon separating the cerebral hemispheres with the hand the
corpus callosum appears intact, and offers fair resistance to
the blade when dividing the hemispheres along this course.
The individual lobes maintain their characteristic forms,
their salient margins, and relative positions; each convolution remains firm, plump, and in close contact with
its neighbour, whilst the arachnoid investment gives a
uniform, smooth, glossy aspect to the consolidated organ
beneath.

A section across the centrum ovale reveals a similar condition—observe the general solidity of the parts; how perfectly relative positions are maintained; the absence of any
gaping of the sulci; observe also that the white substance
cuts with a clean section, and shows no tendency to cling to
the blade. Next open up the lateral ventricles, and observe
the rounded firm aspect of the intraventricular portion of
the corpus striatum and thalamus. Examine the fornix and
septum lucidum—parts especially prone to softening in certain
affections of the brain. Pay special attention to the relative
consistence and size of the corpora quadrigemina, the cerebral
peduncles, the pons and medulla. Now we must be prepared
to meet with all varieties of consistence betwixt the above-
described normal firmness of health and extreme alterations
produced as the result of disease. The consistence, as we
shall illustrate further on, may be so far reduced that the
cerebral substance may be perfectly diffluent, and may be
poured away like thin cream; or, on the other hand, it may
possess the firmness, aspect, and character of the hard-boiled

white of egg ; or even cause the knife to creak in cutting through patches of almost fibro-cartilaginous induration.

Reductions in Consistence.—With the object of rendering ourselves familiar with reductions in consistence, let us consider briefly the appearance and significance of those abnormal conditions of cerebral tissue which lead to softening, dwelling only upon those which are most likely to present themselves frequently to our examination. The conditions best illustrative of reduced consistence are the following :—

> Putrefactive changes.
> Softening as the result of disease.
> > *a*. White softening.
> > *b*. Yellow or gelatinous softening.
> > *c*. Red or inflammatory softening.
> Simple œdematous conditions.

1. **Putrefactive Changes.**—Now these changes in the brain occur sooner or later after death, according to the temperature and humidity of the surrounding atmosphere, modified, of course, by the morbid conditions of its texture and contained fluid. It is on this account that the report of an examination of the brain should invariably be accompanied by a statement of the *number of hours after death* at which the examination was made, the *average temperature of the post-mortem* room, and the *condition of the atmosphere as to humidity*. Fremy, in an elaborate research in 1841 ("Ann. Chim." 2, 463), arrived at the conclusion that the ordinary form of softening of the brain was analogous in its production to the putrefactive process occurring post-mortem ; an opinion which has not been supported by later researches into the chemistry of the brain. We should learn to discriminate between the softening due to this cause, and that due to morbid alteration of texture. We are guided here by observing that the softening is general throughout the encephalon, and attended by much blood-staining from diapedesis, by the evolution of offensive gas, and the presence of the latter

within the blood-vessels and beneath the membranes. We must also take into account the absence of any morbid con-dition tending to produce softening, and the conditions, atmospheric and otherwise, to which the brain has been exposed since death. It is essential, also, to keep in mind the fact that various diseases predispose to early putrefactive changes, whilst the mode of death may expedite or defer the same.

To retard putrefactive changes when the brain cannot be immediately examined, it should be placed in a cold room with a damp cloth thrown over the hemisphere to prevent desiccation of the outer cortical layers; or better still, it may be enclosed within an ice safe. The student can readily extemporize for his own use such a safe, by placing the brain in a jar or box which fits into a larger box, leaving a space of an inch and a half around between the outer and inner box, into which ice is packed. The outer vessels must be thickly and completely enclosed in felt. Such a contrivance he will find of great service in the summer months, when it is difficult to keep anatomical subjects fresh for many hours together. Portions of brain reserved for microscopic examination after hardening should be transferred immediately to methylated spirits or Müller's fluid, or better still to methylated spirits coloured of a dark sherry tint by tincture of iodine. For the fresh methods of examination, we must trust to the ice safe for the preservation of our material, should any delay occur, preservative fluids being scarcely admissible. As regards the latter, however, the strong solution of acetate of ammonium recommended by Sankey [1] is perhaps the best which can be used (sp. gr. 1·040).

2. Indications of Softening.—A *general* reduction in the consistence of the brain will usually reveal itself at first sight by a mere superficial glance. The whole brain assumes an unusually flattened squat appearance, and the hemispheres diverge. The student will find the former due to extension

[1] "New Process for Examining Brain Structure," West Riding Asylum Reports, vol. v. p. 192.

of softening to the central portions of the hemispheres which
support in the normal state the convolutionary surface folded
over them, whilst he observes the commissural band connect-
ing the hemispheres (corpus callosum) tends to split asunder
in a longitudinal direction—i.e., across the course of the great
bulk of its softened fibres. Proceeding to handle the brain,
he finds the convolutions flabby to the touch and presenting
less resistance to pressure—the whole brain is less able to
resist the ordinary force of gravity, and is therefore strongly
contrasted with the plump, erect, and compact aspect of the
healthy organ. Let the student examine the brain of an
advanced general paralytic, and he will find this condition
well represented. Cut across the hemisphere of such a brain
so as to expose the *centrum ovale*, and it will be observed that
the brain-tissue clings to the blade with unusual tenacity
unless the latter be kept constantly wet by water or spirit.
As a result, the cerebral tissues, both grey and white, are
lacerated and tear away in shreds, leaving an unmistakable
softened, rottened aspect of the surface of the section. This
tearing away of brain-tissue is apt to give one a false idea of
the *comparative coarseness of texture*; thus, in the brain of a
general paralytic the adherent, thickened membrane and
softened cortex cannot be cut without leaving a coarse
irregular surface, unless the precaution be taken of using a
very sharp blade, and keeping its surface constantly wet.
Let us now examine more carefully the special form of softening known as white softening.

3. **White Softening of the Brain.**—We may meet
with this condition as one generally diffused through a hemisphere or in patches limited to a convolution, the area of
softening being often no larger than a pea, although more
extensive tracts are usually involved.

In the first place, suppose we have before us a brain in
which the greater part has thus suffered. The affected
hemisphere presents to the touch a soft, boggy feel, sometimes
communicating an almost tremulous, fluctuating sensation to

D

the fingers if the disintegration be extreme. The brain will
be with difficulty removed, the disintegrated tissue tending
to burst through the softened walls which confine it, and
more especially is this the case in the neighbourhood of the
anterior perforated space and commencement of the Sylvian
fissure. All unnecessary manipulation of such a brain should
be avoided, as it is attended with danger to the textural
continuity of internal parts, which should also be examined
at as early a stage as possible. Observe the lateral divergence
of the hemispheres and their flattened aspect above. By
gently drawing the hemispheres apart and introducing a
blade into the longitudinal fissure, a section may be made
outwards across the affected hemisphere, and thus removing
the vault we find the centrum semi-ovale occupied by a mass
of broken-down medulla—all normal cohesion being destroyed
by the presence of a large quantity of interstitial serous
effusion. The disintegrated substance will probably have
the appearance of soft watery or creamy pap, readily raised
upon the blade, or it may be absolutely diffluent, flowing
away as soon as the resistance offered by the confining grey
cortex is overcome. It may, however, retain a certain amount
of cohesion, yet break up readily and wash away on placing
it beneath a stream of water. The ganglia at the base may
be implicated, and the septum pellucidum, grey commissure,
fornix, and corpus callosum will generally be found softened
or even broken down, the ventricle containing more or less
serous fluid. Now such a brain as we have just been considering will be met with in acute hydrocephalus, in thrombosis, or in embolic plugging of the main cerebral arteries.
With such a case before him the student should pay attention
to the following points:—

 a. Degree of softening—
  Does it resist a stream of water?
  Does it render a stream of water milky or break
   down readily under it?
  Is it from the first diffluent?
 b. Extent of area involved.

c. Œdematous condition of the parts involved.

d. Turbidity or clearness and colour of effused fluid in the ventricles.

In noting the *degree of softening* learn to appreciate how very short a period is requisite, after arrest of blood-supply, for the production of extensive nutritive changes.

With what object do we endeavour accurately to map out the *extent* of the destroying lesion ? Our object is to detect that portion of the vascular apparatus involved if the softening be due, as it most frequently is, to plugging of a large branch. This, the more frequent source of extensive white softening, requires, therefore, for its full comprehension, an acquaintance with the *vascular areas of the brain.*

On noting the appearance of the *serous effusion* in the ventricles, so often accompanying white softening, the student will be guided towards concluding as to the presence or not of inflammatory action in the meninges or in the substance of the brain itself. Thus, if the fluid be more or less turbid and discoloured (as the result of a plastic exudate and macerated cerebral substance), we may reasonably suspect an inflammatory condition, and expect to find, upon microscopic examination, exudation corpuscles, compound cells of Gluge, and other products of inflammation. Such are the conditions found in acute hydrocephalus, distinguishing it from a purely non-inflammatory form of hydrocephalus, which occurs where the fluid, though perhaps slightly turbid from broken-down cerebral tissue and shreds from the macerated lining membrane of the ventricles, presents no inflammatory material amongst the débris.

4. Limited Foci of Softening. — Acquaintance with the peculiar arrangement of the cerebral blood-vessels soon leads the student to infer that very minute tracts of softening may result as the effect of thrombosis or embolism, the area involved being dependent upon the size of the clot and the site of its arrest or formation. It is necessary that the mechanism of white softening of cerebral tissue, as a result

of thrombosis or embolism, be thoroughly understood. The
process may be best elucidated by the diagram below.

Fig. 1.—Diagram illustrative of the Effects of Embolic Plugging
(after Rindfleisch).

a a. Portion deprived of its blood-supply by the embolus. A. Artery. V. Vein filled
with blood-clot. The arrows indicate the collateral channels which lead to a hyperæmic
zone around the occluded vessels.

Beyond the obstructed artery is the wedge-shaped area of
its distribution, now anæmic and consequently deprived of
its functional power. Below the embolus are seen swollen
branches, which tend to establish a collateral circulation. If
this fails, we get as a result engorgement of the latter vessels,
and a congestive vascular zone surrounding the wedge-shaped
area. The tissue here becomes swollen and œdematous, and
minute hæmorrhages are apt to occur, whilst the whole
central and peripheral texture becomes broken up by the
effusion, and a true necrosis occurs of the tissue forming the
area of distribution of the nutrient branch which has been
plugged.

All inflammatory foci have, as a result, the production of a
similar congestion and œdema of the surrounding texture,
accompanied by white softening; and as new growths or
tumours are frequently the site of such inflammatory states,
we may find them imbedded in a cavity containing broken-
down tissue.

In addition, therefore, to the appearances which we have to
look for in white softening of the brain, let us also note—

Any obstruction to the vascular supply by thrombi, emboli,

or adventitious products pressing upon the vessels from without.[1]

The establishment of a collateral circulation.

The presence of inflammatory foci, to which the white softening is secondary.

5. **Yellow Softening.** — In the course of pathological studies we frequently meet with appearances which, whilst significant of cerebral softening, differ much from that just described. We find frequently in the white substances of the hemispheres, less frequently in the cerebellum, and rarely on the convolutionary surface of the cerebrum, a focus of softened tissue, varying in size, but scarcely ever larger than a hen's egg, and characterized by its bright straw-yellow colour and soft gelatinous consistence. Pressure causes a yellowish fluid to exude, and when cut across it is found to be somewhat sharply defined from the pale swollen and œdematous tissue around. We recognize in these softened areas the lesion termed "yellow softening of the brain."

Further examination reveals the fact that these spots may be primary, or, on the other hand, secondarily induced around adventitious products, such as tubercle, cancer, cysts, hæmorrhages, or around a patch of inflamed tissue. If secondary to adventitious growths, there is, as Rokitansky points out, an intermediate zone of red softening betwixt the growth and the yellow softened exterior.

It is important that these patches should not mislead the student. He must from the outset regard them as identical in nature with the first form of softening described, the difference in colour being due, according to some, to altered blood-pigment, but according to Rokitansky, to a peculiar chemico-pathological process.

The following facts are to be noted :—

a. The fluid is usually intensely acid.

b. There is little or no vascularity surrounding the patch.

c. Every degree of tint may be found, from white

---

[1] Plastic exudates, tubercle, syphilitic gummata.

softening up to the bright yellow of typical yellow
softening.

d. Colour differs from the rusty or ochre-yellow tint
seen in old apoplectic cavities, etc.

e. Microscopic examination fails to reveal inflammatory
products.

6. **Red or Inflammatory Softening.**—Profound alterations in consistence may thus attend inflammation of the
nervous structures, and we find little difficulty in recognizing
foci of inflammatory softening. We occasionally meet with
portions of acutely inflamed tissue in the medullary strands
of the cerebrum, varying in size from a hazel-nut to that of
an orange, or still larger; yet by far the more frequent site
of such inflammatory foci will be the grey cortex of the cerebrum and the basal ganglia.

Examine a brain which presents a focus of inflammatory
softening in the *medulla* of one of its hemispheres. Note the
following facts:—The *softening of texture* is accompanied by
modifications of *colour* and *general moisture*. Thus the increase
in vascularity gives the affected part a streaky red aspect,
often profusely besprinkled by puncta vasculosa; the œdematous infiltration of texture is recognized, in its *swollen* and
moist aspect. Where the engorgement has reached a high
degree, note the numerous *minute extravasations of blood*, the
streaked or punctated capillary hæmorrhages.

Now all the above characters may be developed in the
course of extreme congestion. What impresses upon the
affected part the stamp of inflammatory action is the presence
of exudates, which modify the above-described appearances
in two directions :—

a. The dark red becomes of a paler and more uniform
hue, whilst exudations of lymph still further modify
the aspect.

b. Alterations in consistence occur; rapid softening or
solutions of textural continuity varying with the
plasticity of the exudates.

Observe, therefore, that the essential features of red inflammatory softening are *profound alterations of consistence*, associated with the pouring out of *inflammatory exudations*.

In advanced stages of red inflammatory softening, the student will, therefore, find the tissues broken down into a reddish pulp, or having a brownish or rusty aspect, washing away freely when held beneath a stream of water. He will also note the results of swelling from congestion and œdema, viz., the flattening of the convolutions of the hemispheres from pressure against the internal table of the skull, and a well-marked prominence of the surface of a section of the inflamed and œdematous textures.

Examine a patch of inflammatory softening in the cortex, and note :—

Inflammatory foci situated in the cortex will implicate by contiguity the superimposed membranes. Observe in these cases that the grey matter is more deeply coloured, and is still more swollen, moist, and softened than when the white matter is involved. The far greater vascularity and the naturally looser texture of the grey matter account for this difference. Next direct attention to the state of the membranes. The pia mater cannot be removed without peeling off with it a layer of the softened cortex, whilst the meshes of the membrane are found infiltrated with inflammatory products, and are both thickened and œdematous. As regards these apparent adhesions of pia mater, observe that the brain substance may tear away with the removal of the pia mater from two causes :—

1st. Mere unnatural softening and loss of textural cohesion of the grey matter, without any true adhesion to the membranes, as in early inflammatory and congestive stages, etc.

2nd. Actual adhesion, inflammatory in origin, may be the cause of this tearing of the cortex.

How are we to distinguish the lacerable cortex in the former case from the genuine inflammatory adhesions of the membranes? Note, in the former case, the surface left by the irregular shreds torn off is soft, pappy, and uniformly smooth or homogeneous; in the latter case, the surface is studded with numerous perforations—the apertures of perivascular channels from which the blood-vessels have been withdrawn. These channels will be seen much dilated from the prior engorgement of the blood-vessels, and the surface generally has a peculiar worm-eaten appearance.

FALLACY TO BE AVOIDED.—We must not regard all cases of encephalitis as necessarily attended by injection and discoloration of tissue. The larger number of cases, in fact, show little or no discoloration; and even where softening has advanced to an extreme degree, the naked eye fails to appreciate in the uniformly dull white pulpy material its inflammatory origin.

What criterion have we here for determining the nature of the lesion? Resource must be had to the microscope and specific gravity bulbs.

The *microscope* shows us nuclei, nucleoli, pigment, and compound granule-cells amongst broken-down cerebral tissue infiltrated with leucocytes and effused lymph along the course of the blood-vessels.

The *specific gravity test* (p. 63) will indicate an invariable increase in specific weight where *inflammatory exudation* has occurred, no matter whether the altered consistence be great or scarcely appreciable. On the other hand, *non-inflammatory white softening* shows an invariable reduction in specific gravity.

By the freezing methods it will be possible to obtain fair sections through inflammatory patches of greatly reduced consistence, and much valuable information will be thus obtained; but in all cases of extreme softening all that can be hoped for is obtained by raising a little of the creamy pulp upon the scalpel, and transferring it to a slide and using a thin glass cover. Sections through the membranes and cortex, in cases where these are involved, will prove highly instructive, and should never be neglected. They should be studied in the fresh state, and the student become thoroughly familiarized with the morbid appearances ere he attempts to mount stained sections of such lesions for permanent preservation. Coarse blood-vessels passing through the inflamed patch should be drawn out and examined microscopically, noting abnormal conditions of their coats and sheath.

7. Œdematous Conditions of the Brain.—It will be seen from the foregoing remarks that this morbid state of the cerebral tissue accompanies all cases of softening of the brain, whether inflammatory or non-inflammatory in origin ; yet it is thought advisable here to consider it independently as a condition, often the only morbid one recognized by the naked eye, in the brain of the chronic insane. As it is chiefly compensatory in its origin, it will be considered together with that compensatory effusion into the ventricles and beneath the membranes of the brain which so frequently accompanies it. If the cerebral tissue becomes much infiltrated with serum, it is liable to break down the texture of the brain. As before stated, this may proceed to a state of white softening, in which the brain-tissue may become quite diffluent, such as we observe in the neighbourhood of the lateral ventricles in acute hydrocephalus and around inflammatory foci. The more frequent condition met with in the brain of the insane is that of a general moist condition of the white matter, which is soft, almost pasty, clings to the blade, and is apt to tear away in shreds of a dirty white hue. Such is the appearance in cases of senile wasting of the brain. The white medullary strands immediately bordering upon the lateral ventricles will usually be found most implicated, being here in closer contact with the serum contained in these cavities. The lesser degrees of œdema are recognized by a peculiar and notable brilliancy of the white substance when cut across, and is one of the most frequent appearances in the brain of the insane. The student must learn to recognize the *general moisture*, *swollen condition*, alterations in *consistence*, and alterations in appearance from a *brilliant to a dull, dirty white hue*, associated with varying degrees of œdema.

If the freezing microtome be liberally used, he cannot fail to appreciate to the fullest extent the varying degrees of œdema of the cerebral tissue; and this not only because it proves the most troublesome obstacle to a free manipulation of delicate sections, but from the tendency of œdematous brain to freeze into a hard icy solid, differing much from the consistence of healthy brain when frozen. It is impossible to cut through such frozen brains except by a modified process to be alluded to further on. In the hardening processes, also, these brains undergo, by dehydration, enormous shrinking.

The serous effusions beneath the arachnoid and into the ventricles, which have been alluded to as often associated with œdematous conditions of the brain, may occur as the result of—

1. Senile atrophy of the brain (compensatory).
2. Pressure by morbid products on the vessels, as in—
    Tubercular meningitis.
    Syphilitic disease of blood-vessels.
    Tumours.
    Growths, or abscesses of mid-lobe of cerebellum pressing on venæ Galeni.
3. As a gradual accumulation from anæmia and other existing morbid conditions, e.g., chronic phthisis.

Where does the effusion of serum attending œdema of the brain occur?

1. Into sac of the arachnoid, i.e., betwixt the cerebral layer of the arachnoid and the polished inner surface of the dura mater. On removal of the brain, we note this fact by examining the amount accumulated in the occipital fossæ above and below the tentorium.

2. Beneath the cerebral arachnoid. Observe how the latter is floated up by the subjacent fluid where this membrane bridges across the sulci.

3. In the meshes of the pia mater. Note its swollen gelatinous appearance from infiltration with serum.

4. Into the ventricles, more or less distending the lateral

ventricles. Stretching tense the thin arachnoid which extends between the cerebellum and medulla over fourth ventricle.

Before concluding the subject of œdematous conditions of brain and compensatory effusions of serum, we would recommend the student to acquire information upon the following facts in all such cases :—

1. Condition of ependyma (lining membrane of ventricles).
     a. As to healthy aspect. *b.* Granular appearance and feel. *c.* Macerated aspect. *d.* Fibro-cartilaginous plates.
2. Turbidity and coloration of serous fluid.
3. Specific gravity of serous fluid.
4. Reaction to litmus paper.
5. Specific gravity of the brain.

Augmented Consistence.—All cases of induration of the brain should induce us to examine the morbid change in the texture as regards its *degree*, its *distribution*, and *nature*. The cases most likely to present themselves to our notice may be embraced under the following categories :—

1st. General augmented firmness of the whole brain from increase of its connective or neuroglia element.
2nd. Limited but extremely indurated patches or nodules due to morbid growths, especially the carcinomata.
3rd. Limited indurations of grey or white matter due to transformation of inflammatory products.
4th. Sclerosis disseminated or distributed down the motor strands.

The student will observe that the increase in consistence may vary from a scarcely appreciable degree, and due merely to a lessened quantity of water (Rokitansky) up to a condition of leathery callous consistence.

A few words upon the varieties of sclerosis met with in cerebro-spinal centres independent of that general increased

firmness from increase of neuroglia observed in the brain
of many epileptics.

1. Superficial Scleroses.—We should also look for
instances of *partial sclerosis of the grey cortex* occasionally to
be seen in epileptics and imbeciles, and characterized by a
peculiar shallow puckering of the surface of the gyri,
which are somewhat indurated, and have a cauliflower
appearance.

2. Hypertrophy of Neuroglia.—Again we may meet
with instances of hypertrophy of the brain, as it is termed,
which has been found due to extreme increase of the
neuroglia around the medullary strands of the hemispheres.
Such a case would present us with increase in *consistence,
elasticity,* and *volume.* In all cases of general increased
consistence it is imperative that the greatest attention
should be paid to examination of the brain under the
following heads:—

   a. Volumetric measurement.
   b. Absolute weight of brain and its divisions.
   c. Specific gravity of grey and white matter.

3. Cicatricial Formations.—The indurations resulting
from inflammatory action will frequently attract attention.
Such patches of callous tissue may surround foci of softening
of the cortex due to plugging, or the site of old hæmorrhages. Cyst-like cavities, with hard fibroid walls, may thus
be formed deep in the substance of the hemispheres.
Sections through such structures should be made for microscopic examination, as they illustrate well the various stages
of transformation of inflammatory products into callous
cicatricial tissue.

4. Disseminated Sclerosis.—If, in slicing a brain,
greyish nodules are found scattered irregularly through the
white substance resembling the cineritious substance in colour,

but becoming of a rosy hue on exposure, and exuding a colourless fluid, the student has probably to deal with a case of disseminated sclerosis. Note in such a case the increased consistence of the sclerosed patches, their vascularity, irregular distribution through the medullary substance, and the comparative immunity from this lesion enjoyed by the cortex. Sections should be preserved for microscopic examination. The extent of implication of the ganglia, pons, medulla, and spinal cord should also be examined.

5. **Descending Sclerosis.**—If again a greyish induration should be met with involving the white strands running down from the motor region of the brain to the internal capsule, or from the latter down into the crura, and on into the lateral columns of the cord, we are dealing with a secondary descending sclerosis (fasciculated sclerosis of Charcot).

§ COLOUR.

The alternating lighter and darker shadings of the cortical layers vary much in the depth of their hue, rendering the distinctness or differentiation of the individual layers more or less apparent. The variations in hue from the normal standard may be distributed in *mottled patches or laminated zones*, or *uniformly spread throughout the whole cortical envelope*. We should consider the conditions usually associated with alterations in the colour of the cortical layers, and first and most frequent amongst such conditions must be placed *alterations in the blood-supply*. The next most frequent condition met with is *disintegration of nervous tissue*: next to these in their relative order of occurrence we find *morbid deposits*; and lastly, *new growths* in the cortex.

Let us first consider the morbid conditions associated with unusual pallor.

1. Partial Mottling and Laminated Pallor.

—A very slight acquaintance with the morbid appearances found in the brain of the insane will suffice to attract attention to the frequent presence of irregular, oval, or circular patches of pallor, very limited in extent, and of a yellowish grey hue. These patches may extend throughout the whole depth of the grey matter or be limited to the superficial layers alone. Now this condition appears almost invariably attributable to anæmic zones, in which special systems of arterioles are involved. It may, however, be due to granular disintegration of nerve-cells and fatty accumulations in the perivascular sheaths.

A similar patchy pallor will also be frequently met with in the central ganglia, but here it will usually be found associated with fatty degeneration of the nuclei in the walls of the blood-vessels.

A portion of a convolution exhibiting this appearance should be placed on the freezing microtome, with the blotchy aspect uppermost. On freezing it will be observed that the whole depth of cortex assumes a uniform pallor throughout, and sections cut off and floated in water *fail to exhibit the pale patches* if the latter be due to anæmic conditions simply; on the other hand, if due to structural disintegration, the appearance vanishes during the frozen state, only to return rapidly when the section is cut. The process of hardening by chrome salts dissipates all patches of pallor due to anæmic zones, for by both processes the blood-vessels are more or less emptied of their contents, and a uniform tint results.

Apart from the presence of fatty or granular débris, the general pallor of the cortex may be due to great paucity of nerve-cells, whilst the blood-supply being diminished the resulting pallor will be associated with a poor differentiation of layers. This apparent fusion of layers and loss of the sharp boundary lines seen in healthy brain is therefore a fact of some significance; it almost invariably points to general malnutrition and disintegration of nerve-cells. Pallor, however, is not necessarily associated with disintegration of nerve-cells, for the latter condition is frequently attended by vascular injection.

2. General Diffused Pallor.—I cannot do better here than describe the appearances found in the case of a patient exhausted by phthisis, in which the typical *pallor* and *malnutrition* of the nervous centres is always so prominent a feature. Such cases are unfortunately so frequent in our asylums that the student will have little difficulty in procuring such a brain for study.

Upon removal of the extremely thin, blanched skull cap, and reflecting the dura mater, the subjacent membranes were seen separated widely from the surface of the brain, buoyed up by fluid which in part was clear and translucent, in part slightly turbid from films and flakes of lymph. The small arterial branches were not visible, the larger primary branches being alone apparent at the vertex. The venous system was represented by engorged trunks, their minutest radicles being well seen. The arachnoid where it crossed the sulci was more or less cloudy, the milky opalescence being due to interstitial change and lymph deposits. In the postero-parietal regions, where the membranes were floated by the distending fluid off from the surface of the gyri, the smaller veins could be traced meandering through the fluid, and dipping down to enter the cortex. Tracing the course taken by these vessels, the smallest radicles were seen emerging from the cortex, uniting here and there with larger twigs, which ramify through the serous fluid and cross obliquely or directly over the summit of the convolution, to terminate in the primary veins which are formed by their convergence. These latter veins run up the sulci on either side of the convolution, superficially exposed and directed towards the median line or longitudinal fissure, where they terminate in the longitudinal sinus. To render their distribution more apparent, we have only to compress the orifice of one of the larger superficial veins near the median line, whilst with the handle of the scalpel placed on a distal part of the same vein, we include numerous converging

radicles betwixt the occluded points. Upon gradually drawing the handle of the scalpel along the course of the vein
upwards, the blood is projected into these radicles, and innu-
merable branchlets before invisible strike across the field over
the convolution, or meander in tortuous arborescent forms
through the subjacent fluid. The student will learn from the
above examination the following relevant and useful facts :—

    1st. The larger arteries being at the base, the smaller
          branches are alone seen near the vertex, and
          these usually deeply seated in the sulci.

    2nd. All the larger blood-vessels usually exposed at the
          vertex are veins.

    3rd. The direction taken by the minute venous radicles
          will be accurately learnt.

    4th. A high degree of vascularity simulating passive con-
          gestion may be induced *artificially* with ease by
          slight force.

It is necessary to observe here that great caution should be
taken by the student against hastily arriving at any conclusion
as to the presence of congestion of the brain from a superficial view of the veins at the vertex—upon this point the
tyro constantly errs. " We must make it a rule to consider
hyperæmia of the cerebral membranes as proved only in those
cases where the finest vessels are also injected, and where the
overloading of the cerebral vessels is not at all in proportion
to the amount of blood in other organs." [1] So says Niemeyer,
reiterating a caution constantly given by pathologists, and as
frequently neglected. Proceeding with our study of the
anæmic brain, the membranes were next stripped and two
facts noted; they were slightly thicker and more tough than
in health, and they were removed with extreme ease. The
student must be familiar with the difficulty experienced in
stripping the membranes from healthy brain, being recommended often to remove them under water. This difficulty
is not alone due to the tenuity of the pia mater and arach-

[1] Niemeyer's " Practical Medicine," translated by Drs. Humphrey and
Hackley, vol. ii. p. 155.

noid, but also to the fact that the former membrane is attached to the surface by the prolongations of connective cells recognized as Deiter's corpuscles. These are by no means numerous in health, yet sufficiently so as to keep the pia mater in firm contact with the cortex. These cells are chiefly found in the neighbourhood of the vessels, where they enter the cortex from the pia mater, and both together they form an impediment to easy stripping. In cases of serous subarachnoid effusion, such as the one under consideration, the membranes have been floated up from the receding surface of the atrophic cortex, and the connections with Deiter's corpuscles have therefore been forced asunder. After noting fully the consistence of the brain according to the plan already given, a section was made exposing the centrum semi-ovale, and the following facts presented themselves:—

1st. The blanched aspect of the cortex and the absence of any reddish striation of the upper layers, such as is seen when the cortical vessels are full of blood. The hue of the cortex was *ashy grey throughout*, except in its most vascular zone, where perhaps the slightest warmth of tint was recognizable, but beyond this, nowhere did any blush suggest the presence of its extremely elaborate vascular apparatus.

2nd. The layers were individually very poorly differentiated.

3rd. The medulla was brilliant white, glistening, slightly reduced in consistence, and showed few or no puncta vasculosa.

We must bear in mind that the vessels of the meninges may be tortuous and turgid with blood whilst the minute vascular supply of the cortex is absolutely diminished, and the layers appear blanched and anæmic. To a certain extent this was the case in the brain we have under consideration. How this occurs (passive or active hyperæmia of the membranes associated with anæmia of the cortex) will be explained in our next section when speaking of hyperæmia of the brain. For the present let us mark well the fact that the *vascular condition of the membranes is no index*, as a rule, to the *state of the cerebral substance.*

E

The brilliant white glistening aspect of the medulla was due, as will be shown further on, to a slight degree of œdema of the brain-substance.

Let us now place a small portion of a parietal convolution of thin and anæmic brain upon our freezing microtome, and cut a fine section through the cortex—float it on the glass slide and drain off superfluous fluid. Take a hand lens and examine first by reflected light.

The cortex is seen as apparently to consist of—

> 1st. An outer light translucent zone extending half way down.
>
> 2nd. A deep opaque white layer extending to the medulla, which is still more opaque and white in aspect.

Raise the slide and examine by transmitted light. We now see—

> 1st. The narrow translucent grey of the first or outer layer of the cortex.
>
> 2nd. A broad less translucent grey zone of the second and third layers.
>
> 3rd. A broad stripe of darker grey extending to the medulla, but separated at one-third its depth by a narrow bluish belt.

We see thus by the naked eye the constitution of the cortex in this region of the brain, except that the narrow second layer cannot be defined from the third layer by *unaided vision*. Let us turn to the cortex prior to section cutting and examine by means of a hand lens under reflected light, and the appearances are as follows:—

> 1st. A grey belt corresponding to the first layer.
>
> 2nd. A broad white belt divided midway by a narrow grey line.
>
> 3rd. A grey belt constituting the deepest layer.

We have, I think, gone sufficiently into the appearances afforded by the anæmic brain for our present purpose. The examination of the various layers above recommended will prepare us not only for recognizing the relative positions of layers which have later on to be minutely studied, but will teach us to estimate roughly—

a. The differentiation of the layers by naked vision.

b. The relative depth of individual layers.

c. The atrophy or normal depth of the cortex.

d. The presence of morbid products.

Of the latter I need here only mention those cases of so-called miliary sclerosis in which the nodules can be distinctly seen in sections on examination by *reflected light*, dotting the surface over with minute opalescent spots. Let us now recapitulate shortly the observations to be made when examining by naked vision a cortex of unusual pallor:—

1st. The extent of pallor throughout the brain.

2nd. Its limitation in depth through the cortex.

3rd. Its disposition in patchy areas or otherwise.

4th. Its association with obscure laminar boundaries.

5th. Its association with pigmentary tints.

6th. Its association with altered consistence and œdema.

7th. Its reappearance in sections from frozen brain.

8th. Examination of sections by transmitted and reflected light.

Redness.—The depth of tint acquired by the grey layers of the cortex depends not alone on the presence of pigmented nerve-cells, but also on the far greater vascularity of the grey as compared with the white matter, whilst the layers most richly supplied with nerve-elements possess also the more abundant capillary supply. The more vascular the layer is, the deeper and warmer will be its tint. Now the capillaries of the cortex are of remarkably fine calibre, and hence, when the ultimate arterioles are injected, the appearance resulting is that of a *uniform rosy blush* more or less dark, the surface *smooth* and *swollen*. This uniform coloration, if at all extreme, is often attended by slight extravasations of blood, and the tissue around presents a different staining from blood-pigment. Such are the conditions found in genuine congestion of the cerebral substance, and the analogous condition is readily recognizable in the soft membranes. The student must make himself thoroughly acquainted with the evidences of a genuine active or passive congestion of the brain, and learn to distinguish it from actual inflammation, where, as already pointed out, the *consistence* as well as *colour* are profoundly implicated, and the *specific gravity also affected.*

1. **Congested Zones.**—One of the most frequent appearances in the cortex of the acute forms of insanity, is a bright arterial zone, which bounds the confines of the white and grey matter of the convolutions, following out accurately the direction of the innermost cortical layer. This congested belt corresponds to the horizontally disposed nexus of blood-vessels into which the larger straight cortical arteries empty themselves after passing through the various cortical layers. Any undue engorgement of the vessels of the pia mater will necessarily affect these larger blood-vessels, and, from their general arrangement, a mechanical element is brought to bear, such that, as I shall endeavour to show further, is significant of a safety-valve action for the cortex in cases of vascular engorgement, relieving the cortex somewhat by their distension from congestive conditions. It is on this account, I believe, that this linear vascularity is so often apparent without a corresponding blush in the cortical layers, indicative of distension of the minute capillaries.

2. **Limited Patchy Congestion of Cortex.**—When, however, the cortex is itself congested, the student will frequently observe irregular bright red patches, suggestive, like the anæmic patches already referred to, of the implication of minute vascular areas, whilst leading down to them are seen the distended straight cortical vessels, giving the upper layers a reddish streaked aspect. The last act of arterial contraction, in which the smaller arterioles have failed to

empty themselves into the venous system, may in part explain this appearance, and caution is necessary, lest we hastily assume that this state is the result of morbid activity. The student should, therefore, examine closely the injected part by the microscope, and look for evidence of minute extravasations, hæmatine crystals, and staining as well as broken-down texture, or other results of inflammatory action and of congestion. This blotchy red aspect of the cortex reappears very frequently in the medulla in similar cases, and is indicative of engorgement of the minute capillaries of these regions.

3. **Puncta Vasculosa.**—It is customary, in judging of the degree of engorgement of the vessels of the white matter, to be guided by the paucity or abundance of the puncta vasculosa, caused by section of the engorged channels—in other words, the number of drops of blood oozing from the divided vessels are supposed to form some criterion of the engorged condition of the vascular system here. It has been shown by Niemeyer that this is no safe criterion, as

1st. The number of puncta vasculosa vary greatly with the fluidity of the blood, which is an important element in their causation.

2nd. They are often almost entirely absent when we have had indisputable evidence of intense vascular engorgement during life.

We must, therefore, be on our guard against attributing too much importance to abundant bloody points on section of the medulla, whilst we should never fail to note whether this condition is or is not associated with distension of the minute capillaries as indicated by fine diffuse patchy redness.

I need not here dwell upon the changes of colour due to inflammatory action, as these have been already disposed of when illustrating anomalies of consistence; but I must here repeat the important caution that the student be not misled by expecting to find engorgement and redness in all cases of encephalitis. The larger number of such cases present no undue red coloration.

Results of Intra-cranial Pressure.—Due allowance must be made for conditions of increased intra-cranial (*but extra-vascular*) pressure. When the cranial cavity is encroached upon by a tumour, abscess, hæmorrhage, or serous accumulation, compensation is made by the outflow of the subarachnoid fluid into the spinal meninges. The *limit of this compensatory arrangement* is soon reached, and then the only available space is that gained at the expense of the *general vascular calibre*. Encroachments upon the cranial cavity therefore eventually empty the blood-vessels, causing partial or more or less general anæmia prior to compression of the brain-substance itself. It is therefore far from improbable that, as Niemeyer teaches, engorgement of the blood-vessels may reach such a degree, that after the limit of subarachnoid compensation has been reached, effusion of serum occurs into the perivascular spaces and brain-tissue, and suffices to compress the minuter blood-vessels and capillaries. Hence we may find in such cases an association of marked cerebral anæmia with tortuous and engorged blood-vessels in the meninges.

Results of Intra-thoracic Pressure.—Venous engorgement of the brain is a frequent result of obstructed circulation through the medium of pleuritic effusions, intra-thoracic growths, fibroid induration, and other changes obstructing or obliterating the vascular system of the lungs. Occasionally, although rarely, there is found an extraordinary engorgement and varicosity of the veins of the pia mater—the vessels winding in all directions, and, as Rokitansky states,[1] even in spirally-twisted coils and intestine-like circumvolutions. In these cases the brain-substance will be found dark and engorged, full of puncta vasculosa and even miliary apoplexies. In one very typical case occurring at West Riding Asylum the brain-surface was actually concealed over extensive tracts by a vast development of varices and contorted vessels, and when cut into, the grey and white substance was

[1] "Pathological Anatomy," Sydenham Soc. vol. iv. pp. 372-3.

not only deeply engorged, but presented very numerous miliary hæmorrhages, forming patches varying from a pea to a florin in extent, and in all degrees from the purely punctiform to the diffuse uniform extravasation. A similar case was exhibited some years back by Dr. Coupland at the Pathological Society of London.[1] In this case the mechanical hyperæmia was due to acute bronchitis—in the Asylum case the immediate cause was the supervention of capillary bronchitis upon fibroid lungs in a patient who gave a history of intemperance. Its association with drink has been recognized by Rokitansky. The occurrence then of such cases should direct the student's attention to the effects of intra-thoracic pressure upon the venous circulation of the brain.

§ VOLUME.

Volumetric Methods for the Brain.—The cubic measurement or volume of the encephalon can be very readily ascertained, and considering the important information it affords, it is a process too frequently neglected in our post-mortem rooms. The volume of the brain is estimated by its displacement of fluid.

1. Dr. Hack Tuke has detailed, in the January number of the *British and Foreign Medical Review* for 1855, the result of examination by this method of sixty-three brains, together with the capacity of the crania. The vessel used by this observer was one of convenient size and shape, with a capacious spout placed at an acute angle with the sides. Water is poured into this vessel up to the level of the spout. Fluid contained within the ventricles and subarachnoid space is allowed first to escape by several long incisions, and then the brain, including the medulla oblongata, is immersed, the displaced water, as it escapes from the spout, being caught and measured, affording an exact criterion of the actual bulk of the brain. The student may, however, prefer the use of a

[1] Reported in the *Lancet* for January 11, 1879.

graduated vessel upon which he can at once read off the displacement of fluid, but the exact graduation of large vessels,
such as would be required, is open to several fallacies; and
these possible sources of error are, of course, wholly avoided
by Dr. Hack Tuke's method.   If such a vessel as that used
by Dr. Tuke be not at hand, a ready method is the one I
have used, as follows :—

2. Half fill with water an inverted bell-glass of sufficient
dimensions, and mark the level of the fluid.   Now immerse
the brain in the fluid, and note the level of displacement,
after which the brain is removed ; and, if the original level
is not exactly maintained owing to imbibition by the brain,
pour in sufficient fluid to compensate for the loss.   Water is
now poured in from a graduated measure up to the displacement level, the amount required for this purpose giving us the
volume of the brain.   The volume should be estimated in
cubic centimeters and cubic inches.

I employ for section-cutting a vessel which answers well
for the volumetric estimate of the brain.   It consists simply
of a large glass vessel, such as is used for the preservation of
brain in museums, but fitted with a stopcock arrangement
in the side near the bottom of the vessel.   It is filled with
sufficient water to cover the brain, the level being marked or
indicated by a weighted float.   The brain is next immersed,
and whilst so immersed, the fluid of displacement is run
off by the stopcock, and measured in a graduated vessel.
Again, in lieu of a graduated vessel, the student may employ
a float, movable along a graduated weighted stem (graduated, of course, for the vessel used).   Such a float should
terminate below in a brass button, which rests against the
bottom of the vessel, keeping the graduated stem in a
vertical position, whilst the float freely moves with the
rising or falling level of the fluid along the graduated stem.

3. When referring to the specific gravity of the brain
an apparatus will be described whereby volume as well as
weight of brain are readily measured.[1]   Such an apparatus

[1] Stevenson's Displacement Apparatus, vide p. 65.

has the recommendation of being cheap and efficient. Whatever arrangement be adopted we should not rest content with the estimate of the volume of the encephalon alone. Each hemisphere should be separately measured, as also the cerebellum, pons, and medulla. It will be found convenient to use smaller and more delicately-graduated vessels for the latter.

Cranial Capacity.—This should be estimated with a view to a comparison between the actual volume and weight of the brain and the dimensions of the cranial cavity, as well as for comparison with the capacity of average-sized skulls. The relationship between cranial capacity and brain-weight has been shown by Dr. Barnard Davis, who has adopted the general rule that a deduction of about fifteen per cent. from the capacity of the cranium gives the " capacity " of the brain, whereby its weight may be readily calculated.[1] The large bulk of cranio-metric observations having been taken upon the dried skull, we should, when dealing with the fresh subject, make allowances for slight divergence ; and, in order to approximate the conditions, the dura mater should be wholly removed from the skull-cap and the base. When, however, our intention is to estimate the difference in any single case between cranial capacity and brain-volume, rather than for more general comparative purposes, it will be necessary to open the skull prior to opening the thorax, so as to avoid emptying the venous sinuses, whilst at the same time the dura mater must not be stripped away from the skull-cap nor the base be uncovered.

1. With the object of estimating the amount of cerebral atrophy, a valuable series of observations were made by Dr. Hack Tuke, and the process he adopted is described as follows : — "The foramina at the base of the brain are

---

[1] " On the Weight of the Brain in the different Races of Man." *Philos. Trans.*, 1868, pp. 506 and 526.

carefully plugged with tenacious clay—that used by statu-
aries for modelling answers best; a small triangular piece
of the frontal bone is removed with the saw; the calvarium
is readjusted to the base, the dura mater being left attached.
The space left by the attrition of the saw in removing the
calvarium is filled with clay; and a narrow bandage with
clay spread upon it is made to surround the cranium three
or four times, covering this space. If this manipulation
has been carefully done, the cavity of the cranium will now
be found as tight as a bottle. Sixty fluid ounces of water
having been measured, a sufficient quantity to fill the cranial
cavity is now poured into it by means of a funnel, through
the orifice in the frontal bone, taking care that the stream
does not wash away the luting of the foramina. The fluid
which remains, after having filled the cranial cavity, is
measured, and being deducted from the sixty ounces gives
the amount employed. To this must be added half an ounce
for the space occupied by the luting." Having thus obtained
the cranial capacity, he deducts from it the brain volume,
and obtains thus the exact measurement of shrinking or
atrophy. Millet seed and sand have been used for measuring the capacity of the cranial cavity, and the latter in the
dried skull has answered admirably. It must, however, be
borne in mind that a fallacy may be introduced by the
employment for this purpose of materials liable to be
influenced by temperature. Fluids are, of course, more
open to this objection, their expansion by heat being pro-
portionately more rapid than solids, and hence it would
appear that sand is preferable to water in these investigations.

2. Mustard seed was employed by Professor Flowers in
his extensive series of researches.[1]

3. The method adopted by myself arose from my employment of solid paraffin in obtaining casts of the brain-

---

[1] The capacity of the crania contained in the Hunterian Museum of the
Royal College of Surgeons was obtained by Professor Flowers in this
manner.

surface, and the interior of the skull. It was soon apparent that the finest and most delicate impressions and most perfect casts could be obtained by the use of this substance, and that it did not share in the great disadvantage which accrues from the use of plaster of Paris, viz., that of great contraction during solidification. It is in this respect also far superior to white wax, which, as is well known, contracts much whilst solidifying. I proceed as follows:—Fill up the foramina at the base as previously described. A triangular or wedge-shaped piece is now sawn out of the occipital bone after removal of the calvarium, but retained *in situ*. The base is then filled up by melted paraffin, the skull-cap replaced and fastened by luting, just as in Dr. Tuke's process, having previously trephined a piece out of the frontal bone. Through the latter orifice more of the paraffin is poured in until the cranial cavity is filled. When cool and solid remove the calvarium, as well as the wedge-shaped piece of bone from the occiput, and then gentle pressure from behind tilts the solid mass out of the cranium, when it will be found to form an exquisite mould of the interior. The mould thus obtained is now to be measured by displacement, whence we obtain the cranial capacity. This method has afforded me great satisfaction, as it gives data of great value at the expense of very little trouble: it supplies us with a method of—

1st. Estimating cranial capacity.

2nd. Gives us an exact mould of the cranial cavity.

3rd. As a permanent record—numerous linear angular measurements and volume measurements may be obtained.

4. From some invaluable measurements by the late lamented Broca,[1] it was found that of 115 skulls of individuals living in the twelfth century, the *average* capacity was 1425·98 cubic centimeters; whilst of 125 skulls of the nineteenth century the average capacity was 1461·53 cubic centimeters. The following table of percentage will be found of

---

[1] Quoted by Vogt, "Lectures on Man." Anthrop. Soc.

use by those interested in the subject of cranial capacity. It is given by Le Bon as illustrative of the relationship of race to cranial capacity :—

CRANIAL CAPACITY IN DIFFERENT HUMAN RACES.

CRANIAL CAPACITY.	MODERN PARISIANS.	PARISIANS OF 12TH CENT.	ANCIENT EGYPTIANS.	NEGROES.	AUSTRALIANS.
Cubic Centimeters.					
1,200 to 1,300 .	0·0	0·0	0·0	7·4	45·0
1,300 to 1,400 .	10·4	7·5	12·1	35·2	25·0
1,400 to 1,500 .	14·3	37·3	42·5	33·4	20·0
1,500 to 1,600 .	46·7	29·8	36·4	14·7	10·0
1,600 to 1,700 .	16·9	20·9	9·0	9·3	0·0
1,700 to 1,800 .	6·5	4·5	0·0	0·0	0·0
1,800 to 1,900 .	5·2	0·0	0·0	0·0	0·0

§ WEIGHT.

Gravimetric Methods.—The student must make himself familiar with the various circumstances which modify brain-weights. Some of these conditions have been summarized by Bastian.[1] They appear to be chiefly as follows :—

1. Length and nature of illness.
2. Mode of death (vascular engorgement favouring high weight).
3. Certain neuroses, as epilepsy.
4. All conditions inducing sclerosis.
5. Atrophy.
6. Congenital micro- or megalo-cephaly.

The student must also take into consideration the relationship between weight of brain and the *age* and *sex* of the individual as well as the *weight and height of body*.

---

[1] "The Brain as an Organ of Mind," chap. xx.

Absolute Weight of Brain.—Before weighing the brain, all fluid from the subarachnoidal space and ventricles must be drained off, and allowance made for the membranes, which will scarcely exceed an ounce in weight. The plan adopted at the West Riding Asylum, however, is to first strip the brain of its membranes, to liberate the fluids in the ventricles, and weigh the whole encephalon. The membranes can afterwards be weighed if thought necessary. In all cases when the pia mater is firmly adherent to the cortex, it is as well to weigh the brain prior to stripping, as large shreds of the cortex are often removed in these cases. Precaution must be taken, however, by incisions, to release any accumulation of fluid in the subarachnoid spaces and meshes of the membrane, and, subsequent to weighing, allowance must be made for the pia mater and arachnoid. After the weight of the encephalon has been obtained, the cerebellum and pons must be removed by dividing the crura cerebri *close to the pons*.[1] A longitudinal incision is then carried through the median line of the corpus callosum from before backwards, so as to separate the two hemispheres. The peduncular connections of the pons and medulla with the cerebellum are then divided, and the former (pons) also separated from the medulla at its natural line of division. This is the method adopted at West Riding Asylum, but for several years it was the custom here to separate the frontal lobes from the remaining posterior part of the brain by an incision carried through the fissure of Rolando. The weights of these individual parts are then taken.

It will be found advisable to use the metric system in all our estimates of capacity, volume, and weight, and all the standard brain-weights afforded by the elaborate tables of Tiedemann, Reid, Boyd, Wagner, and others should be expressed in grammes rather than ounces. Dr. Sharpey, after an elaborate analysis of brain-weights given by Glen-

---

[1] A modification of the plan is recommended in the section on the dissection of the brain as preferable when the question of weight is of secondary import (p. 69).

dinning, Sims, Tiedemann, and Reid, supplies us with the
following valuable results :—[1]

			ozs.
Maximum weight of adult Male Brain			65
Average	„	„ „	49½
Minimum	„	„ „	34
Maximum	„	„ Female Brain	56
Average	„	„ „	44
Minimum	„	„ „	31

The heaviest human brain on record has been described by
Dr. Morris,[2] and was carefully examined at University
College Hospital : the weight was 67 oz.

Compatible with ordinary intelligence, the lowest limit of
the human brain, as regards weight, is, according to Gratiolet,
900 grammes, and, according to Broca, 907 for the female,
and 1049 grammes for the male.[3]

Specific Gravity.—Researches upon the specific gravity
of the encephalon are becoming more and more interesting
with our knowledge of cerebral localization and intimate brain-
structure. The late suggestive work of Danilewsky upon the
relative amount of grey and white matter in the brain depends
greatly upon the accuracy of the specific gravity for its cor-
rectness. The more important methods adopted are those of
Drs. Bucknill, Sankey, and Peacock.

[1] "Elements of Anatomy," 7th Edition, vol. ii. p. 568.
[2] *Brit. Med. Journ.* Oct. 26, 1872, p. 465.
[3] Quoted by Bastian, "The Brain as an Organ of Mind," 1880, p. 365.
In connection with the weight of the brain in the insane, the student
will find most valuable information in articles by Dr. Crochley Clapham,
contained in the 3rd and 6th vols. of the "West Riding Asylum Medical
Reports." His observations embrace 1,200 cases of insanity. Further
information upon this subject may be obtained by reference to the
following :—Sims, *Medico-Chirurg. Trans.* vol. xix. ; Glendinning, *Medico-
Chirurg. Trans.* vol. xxi. ; Tiedemann, "Das Hirn des Negers," Heidelberg,
1837 ; Reid, *London and Edin. Month. Journ. Med. Science,* April, 1843 ;
Thurman, *Journ. Mental Science,* 1866 ; Wagner, "Vorstudien," 1862, 2ᵉ Abh.
pp. 93-95 ; Peacock, *Month. Journ. Med. Science,* 1847, and *Journ. of Pathol.
Soc.* 1860 ; Boyd, *Philos. Trans.* 1860 ; Bastian, "The Brain as an Organ
of Mind," 1880 ; Clapham, loc. cit. and *Journ. of the Anthropolog. Inst.*
vol. vii. p. 90.

1. **Bucknill's Method.**—Dr. Bucknill was the first who originated a satisfactory and ready method for estimating the specific weight of brain. The following is his detailed account of the process :—" The specific gravity of the cerebrum and cerebellum is ascertained by immersing a portion of each in a jar of water wherein a sufficient quantity of sulphate of magnesia has been dissolved to raise the density of the fluid to the point required, adding water or a strong solution of the salt, until the cerebral mass hangs suspended in the fluid without any tendency to float or sink, and then, by testing with the hydrometer, the specific gravity is thus found with great delicacy and facility, a difference of half a degree in the density of the fluid being indicated by the rise or fall of the substance immersed. The soluble salt is chosen, for its possessing no astringent or condensing action upon animal tissues."[1]

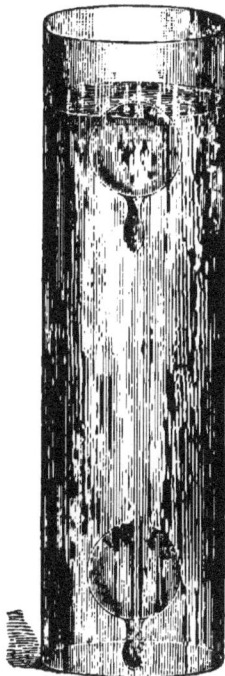

2. **Sankey's Method.**—This process is the one which has been used most extensively by Dr. Crichton Browne at West Riding Asylum, and it is the method which I have myself invariably adopted. It appears to me in every respect highly satisfactory and simple. We require a series of cylindrical glass jars, such as the one figured, and a set of graduated hollow glass bulbs, which can now be readily obtained.[2]

FIG. 2.
SPECIFIC GRAVITY TEST.

These glass bulbs are accurately graduated, or rather marked with the specific gravity of the fluid in which they

[1] *Lancet*, 1852, vol ii. p. 589.
[2] Such a set of graduated bulbs may be obtained of Mr. Stevenson, Philosophical Instrument Maker, No. 9, Forrest Road, Edinburgh.

would hang suspended when immersed, neither tending to float nor sink. The glass jars are partly filled with water, and then a concentrated solution of Epsom salts added to each, until its specific gravity is such that whilst one bulb floats, a bulb two degrees higher sinks. Thus a series of jars are filled as represented below :—

Bulbs floating	1030	1031	1032	1033	1034
Specific gravity of fluid =	1031	1032	1033	1034	1035
Bulbs sunk	1032	1033	1034	1035	1036

The middle line represents the specific gravity of the fluid, and therefore of the portion of brain which tends neither to float nor sink, but to remain suspended wherever placed. It will occur to the student that the jars may be so graduated, that whilst one bulb floats, another, *one degree* higher, sinks, and hence that half a degree specific gravity may be indicated as follows :—

Bulb floats	1030	1031
Specific gravity of fluid and suspended body	1030·5	1031·5
Bulb sinks	1031	1032

This is a degree of nicety, however, to which the use of the beads should not be pressed, and if required, should demand in preference the use of the 1000-gramme specific gravity bottle. A series of jars should be graduated from 1028 to 1050, so as to enable us to deal with white or grey matter of cerebrum, cerebellum, and central ganglia. Occasionally still more dilute solutions will be requisite.

A minute piece of white or grey matter is now raised by a scalpel and placed upon the perforated spoon or scoop, which is gently lowered into one of the jars of saline solutions, and the fragment of brain turned off the scoop and closely observed. If it sinks, it is of course of higher specific gravity than the fluid, and must be passed on to stronger solutions until, reaching one of its own density, the fragment remains stationary where placed. If, on the other hand, it floats, it must, of course, be moved to a solution of less specific weight, until the same conditions are obtained. Should the specific

gravity beads be chosen by the student for this purpose, it
would be well for him to attend to the following rules :—

a. After obtaining the specific gravity of any one portion
of brain, repeat the trial with fresh portions at least two or
three times, so as to ensure perfect accuracy.

b. Examine the brain as soon as possible after death,
wholly rejecting such as show the least evidence of commencing decomposition.

c. Keep the jars covered with glass squares in a cool room,
so as to avoid dust and evaporation as far as possible.

d. Before commencing any series of observations, note the
position of the bulbs, any tendency of the lower bulbs to float
from evaporation of the solution being rectified by the addition
of water.

3. Stevenson's Displacement Apparatus.—A simple
and most reliable apparatus has been devised by Mr. Steven-

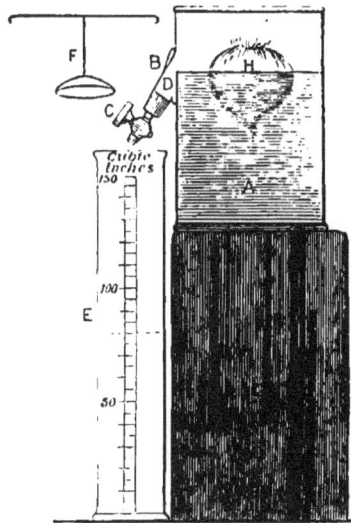

FIG. 3.—STEVENSON'S DISPLACEMENT APPARATUS.

son for estimating at the same time both the bulk and specific
gravity of large irregular bodies by the principle of displace-

F

ment.   The apparatus consists of a jar, A, fitted with a
large drooping tubulature, B, with stopcock attached.   A
metal gauge, D, having a thin horizontal straight edge inside
the jar, determines the exact water-level, and over it the
surplus water flows.   The other jar, E, is graduated on the
sides into cubic inches, avoirdupois lbs. and ozs., and thousandth
parts of a gallon.   The packing case, G, serves as a stand for
the jar.   In taking the specific gravity of the brain, or any
body a *little* heavier than water, the jar must be filled with a
solution of salt or sulphate of magnesia up to the level of
the gauge, the stopcock being closed.   The brain is then
immersed in the fluid, in which it ought to float, and the fluid
immediately rises in the jar; let it come to rest; open the
stopcock, and the quantity displaced will flow into the
graduated jar and represent the *weight-indication* of the brain
(since any floating body displaces exactly its own weight of
the fluid).   The stopcock is now shut, and the brain depressed
by the brass piece, F; and when the water has again come
to rest, open the stopcock and receive the further displacement
in the jar E.   The whole bulk of displaced fluid read off the
graduated jar indicates, of course, the volume of the brain.
To obtain the specific gravity, multiply the weight-indication
by the specific gravity of the fluid and divide by the volume-
indication.   I need scarcely indicate to the student the
simplicity and value of this method, whereby the *weight*, the
volume, and the *specific gravity* of the brain may be simultaneously obtained.[1]   The apparatus, as figured, is somewhat
smaller than what would be required for brain.

In estimating the specific gravity of the encephalon, the
student will soon appreciate the fact that the specific weight
varies not only in the cerebrum, central ganglia, and cerebellum, but that variations occur over different regions of
the cerebrum, suggesting the importance of a comparative
examination by this method of the various cerebral convolutions.   By cautious manipulation he will also be able to
show that the specific gravity of the cortex varies with its

[1] This apparatus is sold by Mr. Stevenson, 9, Forrest Road, Edinburgh.

depth. Dr. Sankey observes that the specific gravity diminishes in the ratio of ·001 for every twenty-four hours after death. This must, however, be greatly modified by the temperature of the surrounding atmosphere, and other conditions favouring putrefactive changes. The following embrace the more important results obtained by different observers, and they will serve as a useful guide to the student in his prosecution of similar observations [1] :—

Average specific gravity of whole				Encephalon	1036	(Bucknill).
"	"	"	"	Cerebrum	1030–'48	(Aitkin).
"	"	"	"	Cerebellum	1038–'49	(Aitkin).
"	"	"	"	Grey matter	1034	(Sankey).
"	"	"	"	White matter	1141	(Sankey).
"	"	"	"	Central ganglia	1040–'47	(Aitkin).

Proportion of White and Grey Matter in the Brain.

—The specific gravity of the grey and white matter, together with that of the whole brain, has been applied to solve the very important problem of the relative percentage of grey and white matter in the human brain. Danilewsky has lately published his results and method of procedure, an account of which may be seen in the *Centralblatt f. d. Med. Wissenchaften*, No. 14, 1880.[2] The formula given by him is as follows :—

$$x = \frac{Pb\,(p - a)}{p\,(b - a)}$$ where x is the *quantity* of grey or of white matter, p the *specific gravity* of the whole brain, a of the grey and b of the white substance, and P is the *weight* of the whole brain. He gives, as the results of one series of experiments, a percentage for the grey matter of 37·7 to 39, against a percentage of 60·3 up to 61 for the white substance. By taking the average depth of grey matter it is, of course, an

---

[1] References to work done in the Specific Gravity of the Brain :—Bucknill. *Lancet*, vol. ii. 1852, and the *Med.-Chir. Rev.* 1855; also "Psychological Medicine," Bucknill and Tuke, 3rd edit. pp. 520 and 587; Sankey, *Brit. and For. Med.-Chir. Rev.* 1853; Peacock, "Trans. Pathol. Soc. of London," 1861-2; Aitkin, "Science and Practice of Medicine," 1865, vol. i. p. 265.

[2] Vide Abstract by Dr. Geoghegan, *Journ. Mental Science*, p. 437, Oct. 1880.

easy matter to obtain the surface measurement of the brain.
Wagner's researches upon the superficial area of the brain are
of interest here.[1] The student, however, must be warned
against regarding results so obtained as other than merely
approximate, as it is next to impossible to exclude the
numerous fallacies to which the estimation of average specific
gravity is liable. Thus the specific gravity of the grey
matter is known to vary with its *depth*, with its *local distribution or area*, with the amount of *vascularity* of the tissue, and
time after death. Beyond this the very cases in which the
pathologist would be interested in estimating the relative
proportions of grey and white matter are those which are
subject to such diffuse and local change in consistence that
an average specific gravity for either white or grey could not
be attained with any degree of exactness. For comparative
investigation of healthy brain, the process adopted by
Danilewsky promises to afford valuable results as long as
the greatest care is observed to exclude fallacies.

[1] Quoted in Quain's "Anatomy," vol. ii.

§ EXTERNAL ASPECT OF THE BRAIN.

It may be useful here to review the course to be pursued in removing and preparing the various parts of the brain for coarse examination, and the following method is recommended as in every way the more reliable and satisfactory one.

In the first place the head should be so supported as to command a good light and a full and satisfactory view of the brain *in situ*.

In the next place never neglect the rule of opening the skull and examining the brain *prior to opening the thorax* and dividing the great vessels.

Cranial Membranes and External Aspect of Brain. —The skull-cap having been removed *see. art.* we proceed as follows :—

1. Examine upper surface of dura mater in its relationships to the skull (pp. 3–5).

2. With a forceps raise a fold of dura mater anteriorly, run the scalpel through it, and carry the blade along either side on a level with the sawn edge of the cranium far back towards the occiput, stopping short of the middle line.

3. Reflect these lateral halves of the membrane towards the middle line, observing the condition of the large superficial veins running into the longitudinal sinus.

4. Replace the dura mater, and with a curved bistoury

open up the longitudinal sinus in its whole length back towards the torcular herophili, noting the condition of the parts (p. 5).

5. Divide the connection between dura mater and crista galli, and seizing the anterior end of the falx cerebri, forcibly draw the membrane backwards, dividing the junction of the superficial veins in this course, and thus expose the surface of the hemispheres.

6. With a good light thrown on the subject, and, if requisite, the aid of a hand lens, observe the appearances presented at the vertex, noting the condition of the vessels, membranes, and general conformation of the brain. It is absolutely necessary to pay attention to this point, as the appearances presented *in situ* are often greatly modified or wholly lost upon removing the brain at a later stage.

7. Gently raise the tips of the frontal lobes from the orbital plates, carefully removing the olfactory bulbs with the brain, and then using a little gentle traction, the optic nerves are exposed and divided close to their foramina. With the same blunt-pointed curved bistoury divide successively the infundibulum, carotid arteries, and third nerve. A good view is thus afforded of the tentorium, which should be divided along its attachment to the ridge of the temporal bone, dividing at the same time the fourth nerve. The base of the brain is now exposed, and the fifth, sixth, and seventh nerves readily divided, after which the blade is passed down into the vertebral canal on either side of the medulla, and, cutting forwards, it severs the vertebral arteries, the eighth and ninth pairs, and spinal accessory nerves. A sweep of the scalpel across the front of the cord liberates the brain, which can be now raised out of the cranial cavity with ease.

8. The next procedure is to place the brain base upwards in the skull-cap, the latter being conveniently steadied by any simple contrivance. This enables us with less sacrifice to the appearances at the vertex, to study the important region of the base. In doing so, follow out the instructions

already given (p. 15), taking the parts in the following
order :—

    a. Arrangement and appearance of the great vessels at
the base.

    b. Condition of the membranes and subjacent gyri.

    c. Condition of the various cranial nerves.

    d. Open up the fissure of Sylvius on either side, tracing
the large arterial branches upwards; also examine
nutrient supply to the basal ganglia at anterior and
posterior perforated spots.

    e. Strip the base of its membranes, noting adhesions,
etc. etc.

9. The cerebrum may now be turned out upon the dissecting tray, with vertex uppermost, and the membranes
carefully examined and stripped. Take note of the arrangement of the gyri, the presence of superficial lesions, general
consistence, and colour.

10. The dura mater and its sinuses at the base of the
skull may be now examined (p. 8), completing thus the
examination of the cranial membranes and the external
aspect of the brain.

§ EXAMINATION OF INTERNAL STRUCTURE OF THE BRAIN.

1. Introduce a large section knife into the longitudinal
fissure, and cut outwards across each hemisphere, about ¾ inch
above the corpus callosum, exposing the *centrum ovale minus.*
Repeat the same procedure on a level with the corpus
callosum, exposing the greater centrum ovale, and exhibiting
the continuation of the transverse strands of this great commissural tract with the medulla of the hemisphere. The
student will now proceed to note the relative and absolute
amount of grey and white matter, and inquire into the consistence, colour, and other physical qualities of these parts in
the manner already detailed (Chap. IV.).

2. By an incision carried from before backwards across the fibres of the corpus callosum, a short distance on each side of the median line, the lateral ventricles are opened. To expose the structures more fully, divide the corpus callosum anteriorly, and whilst gently reflecting it cut backwards, with a scissors, through the vertical septum (septum lucidum), which is attached to its under surface, and which, descending, divides the lateral ventricles into two cavities. After reflecting the corpus callosum, and dividing it across behind, note the condition of the following structures *seriatim :*—

> *a.* The caudate nuclei or intra-ventricular portions of the corpora striata.
> *b.* Septum lucidum, enclosing betwixt its walls the fifth ventricle.
> *c.* Exposed anterior tubercle of the thalamus opticus.
> *d.* Stria cornea, or exposed part of the tænia semi-circularis, coursing between the thalamus and corpus striatum.
> *e.* Fringed margins of the velum interpositum or choroid plexus of lateral ventricles.
> *f.* Fornix, and behind its descending pillars, the foramen of Monro.
> *g.* General condition of lining membrane or ependyma.

3. Divide the fornix anteriorly close to its descending pillars, and after examining the subjacent velum interpositum, reflect both structures backwards, exposing the thalami optici, separated by the intervening third ventricle, and note the condition of the following parts :—

> *a.* Grey matter forming the boundaries of the central cavities.
> *b.* Anterior, middle, and posterior commissure.
> *c.* Pineal gland, and its ganglionic extensions, or peduncles.
> *d.* Corpora quadrigemina (nates and testes).
> *e.* Follow out the fornix posteriorly into the descending cornua of the ventricles.

4. On separating the occipital lobes from the cerebellum the curved bistoury may, by a little careful dissection, be made to cut through both *crura cerebri* obliquely upwards, to meet *above the corpora quadrigemina* at an obtuse angle in the third ventricle in front of the posterior commissure.

By so doing we separate from the cerebrum the cerebellum, pons, and medulla, together with the *corpora quadrigemina* and *pineal gland intact*. This method is certainly preferable to dividing the crus at the level of the superior cerebellar peduncle, which is so often done at the sacrifice of the normal relationships of most important regions.

5. Separate the cerebral hemispheres by median section through the slight remaining connections, and proceed as follows :—

a. Slice one of the hemispheres in a direction parallel to the sections already made when exposing the centrum ovale. Let numerous sections in this direction be made at different planes from above downwards through the whole hemisphere, so as to expose the structure of the thalamus and corpus striatum down to the base.

b. Slice the remaining hemisphere in a direction at right angles to its long axis, or, in other words, from above downwards, beginning our sections near the olfactory bulb, and proceeding as far back as the occipital lobe.

These sections (*a* and *b*) will familiarize the student with the *relationships and coarse structure* of the basal ganglia— points of considerable moment. Examine, therefore, by aid of these sections, the medullary tracts forming the *internal and external capsules*, the *lenticular and caudate nucleus* of the corpus striatum, the *thalamus opticus* and its various regions and environment. All these parts may be studied in succession, referring to good illustrations, such as appear in the last edition of Quain's "Anatomy." The intimate structure and relationships of these regions will be dealt with further on.

6. Remove the cerebellum from its attachments by dividing the three peduncles, and proceed to examine successively—

 a. Superficial aspect of membranes covering pons and medulla.

 b. Condition of blood-vessels and superficial origin of the cranial nerves.

 c. Condition of valve of Vieussens, aqueduct, and fourth ventricle.

 d. By transverse sections expose structure of nates and testes, of the pons, its anterior, or motor, and posterior or tegmental tract, of the medulla oblongata, noting the relative dimensions and appearance of the olivary bodies, the pyramidal and restiform tracts.

7. Lastly, after examining the lobules and membranes of the cerebellum, we divide it into an upper and lower half by an incision carried through both hemispheres from behind forwards, exposing the central medulla.

Vertical section—*i.e.*, at right angles to the lamina of the cerebellum—will display its foliated arrangement, and betwixt vermiform process and the middle of each hemisphere we pass through the plicated corpus dentatum.

CHAPTER VI.

SINCE the prosecution of researches into the minute anatomy of the brain will necessarily lead us into structural details of infinite diversity (for the complete histology of the human cortex cerebri is infinitely more complex than any of our classical works on this subject would lead us to conceive), we shall have to bestow a predominating share of our attention upon cerebral localization. Hence at the outset it is essential that we obtain clear and definite ideas of the various sub-divisions of the brain, both as regards convolutionary arrangement and central medullated fasciculi. A knowledge of the convolutions and sulci must be immediately acquired, and from this superficial topography we learn readily to appreciate the relative positions of those smaller areas which, from experimental data, have been proved to possess diversi-fied functional endowments, as well as those cortical tracts which histological inquiry has invested with structural peculiarities.

It is not within the scope of this manual to give even an outline of the internal medullated structure of the brain, but it may prove of service here to indicate Professor Turner's nomenclature of the convolutions which, from its great simplicity and clearness, has found such general favour.[1] For minute details as to the course, relationships, and devia-tions of these gyri, the student is referred to that very

---

[1] "The Convolutions of the Human Cerebrum Topographically Con-sidered," by William Turner, M.B. Lond., F.R.S.E. *Edin. Med. Journ.* June, 1866.

excellent translation of Ecker, " On the Convolutions of the Human Brain," by J. C. Galton. From this work, with the editor's kind permission, I have tabulated the various English and foreign synonyms, as likely to prove of value in the further prosecution of our studies in this department. The photographs from Bischoff's work on the Convolutions are intended for immediate reference in the laboratory.

GENERAL REFERENCE TO THE PHOTOGRAPHS.

I. Frontal Lobe.   II. Parietal Lobe.   III. Occipital Lobe.
IV. Temporo-sphenoidal Lobe.

1. Superior Frontal Gyrus ..   } Separated by the Supero-frontal
2. Middle Frontal Gyrus.. .. }   Sulcus.

3. Inferior Frontal Gyrus ..   { Separated by the Infero-frontal
                       {   Sulcus.

4. External Orbital Gyrus.
6. Ascending Frontal Gyrus .. } Separated by the Fissure of
7. Ascending Parietal Gyrus .. }   Rolando.

9. Postero-parietal Lobule .. }
11. Supra-marginal Lobule .. } Separated by the Intra-parietal
12. Angular Gyrus.. .. .. }   Fissure.
13. Third and Fourth Annectant Gyri.
14. First Annectant Gyrus.

10. Quadrilateral Lobule .. } Separated by the Internal Pa-
16. Cuneus, or Superior Occipital } rieto-occipital Sulcus.
      Lobule.. .. .. .. }

*17. Lingual Gyrus .. .. .. } Separated by the Calcarine
                        } Fissure.

*18. Fusiform Gyrus .. .. } Separated by the Collateral
                        } Fissure.

19. Superior Temporo-sphenoidal Gyrus } Separated by the Parallel
20. Middle Temporo-sphenoidal Gyrus } Fissure.
21. Inferior Temporo-sphenoidal Gyrus } Separated by the Collateral
22. Uncinate Gyrus .. .. } Fissure.

5. Median Aspect of Superior Frontal }
    Gyrus .. .. } Separated by the Calloso-mar-
23. Convolution of the Corpus Cal- } ginal Fissure.
     losum .. .. }

N.B.—At the Base No. 2, 3, 4, indicate respectively the Internal, Posterior, and External Orbital Gyri. I need scarcely add that the darker and lighter shading of the brain, as shown in the photographs, is well calculated to bring into strong relief the more distinctive areas and divisions of the surface. The relative areas occupied by different lobes over each aspect of the hemisphere are thus better displayed than by outline figures, whilst the distinctive character of the gyri is not interfered with.

* Neither of these terms is included in Turner's nomenclature. They are adopted by Ecker.

§ LIST OF SYNONYMS FROM ECKER.

ADAPTED TO TURNER'S NOMENCLATURE.

SUPERIOR FRONTAL GYRUS.

Erste oder obere Stirnwindung	(*Ecker.*)
Étage frontal supérieur ou troisième .. }	(*Gratiolet.*)
Pli de la zone externe .. .. }	
Supero-frontal gyrus	(*Huxley.*)

MIDDLE FRONTAL GYRUS.

Zweite oder mittlere Stirnwindung	(*Ecker.*)
Étage frontal moyen	(*Gratiolet.*)
Medio-frontal gyrus	(*Huxley.*)

INFERIOR FRONTAL GYRUS.

Dritte oder untere Stirnwindung	(*Ecker.*)
Pli frontal inférieur ou premier ou étage	
surcilier	(*Gratiolet.*)
Infero-frontal gyrus	(*Huxley.*)

ASCENDING FRONTAL GYRUS.

Vordere Centralwindung	(*Huschke.*)
Processi cuteroidei verticali di mezzo (anterior	
part)	(*Rolando.*)
Circonvolution transverse pariétale antérieur	(*Foville.*)
Premier pli ascendant	(*Gratiolet.*)
Antero-parietal gyrus	(*Huxley.*)

SUPERIOR TEMPORO-SPHENOIDAL GYRUS.

Erste obere Schläfenwindung	(*Wagner.*)
Antero-temporal	(*Huxley.*)
Pli temporal supérieur ou.. .. .. }	(*Gratiolet.*)
Pli marginal postérieur .. .. .. }	
Gyrus temporalis superior sive infra marginalis	(*Huschke.*)

MIDDLE TEMPORO-SPHENOIDAL GYRUS.

Zweite oder mittlere Schläfenwindung ..	(*Wagner.*)
Mittlere Schläfenwindung	(*Huschke.*)

Pli temporal moyen ou partie descendante du
pli courbe .. .. (*Gratiolet.*)
Medio-temporal gyrus .. .. (*Huxley.*)

INFERIOR TEMPORO-SPHENOIDAL GYRUS.

Dritte oder untere Schläfenlappenwindung }
Gyrus temporalis tertius sive inferior .. } (*Wagner.*)

LINGUAL LOBULE.

Untere innere Hinterhauptwindungsgruppe (*Bischoff.*)
Lobulus Lingualis—Zungenläppchen .. (*Huschke.*)
Gyrus occipito-temporalis medialis .. .. (*Pansch and Ecker.*)

FUSIFORM LOBULE.

Untere aussere Hinterhauptwindungszug.. (*Bischoff.*)
Gyrus occipito-temporalis lateralis .. .. (*Pansch and Ecker.*)
Spindelläppchen .. .. (*Ecker.*)
Lobulus fusiformis—Spindelformiges
Läppchen .. .. (*Huschke.*)

ASCENDING PARIETAL GYRUS.

Hintere Centralwindung .. .. .. (*Ecker.*)
Processi enteroidei verticale di mezzo
(posterior segment).. .. .. .. (*Rolando.*)
Circonvolution tranverse medio-pariétale .. (*Foville.*)
Deuxième pli ascendant .. .. .. (*Gratiolet.*)
Postero-parietal gyrus .. .. .. (*Huxley.*)

POSTERO-PARIETAL LOBULE.

Oberes Scheitelläppchen .. .. .. (*Ecker.*)
Gyrus parietalis superior .. .. .. (*Pansch.*)
Lobule du deuxième pli ascendant .. .. (*Gratiolet.*)
Erste Scheitellappenwindung .. .. (*Wagner.*)
Obere Scheitelbeinlappen .. .. .. (*Huschke.*)
Obere innere Scheitelgruppe .. .. .. (*Bischoff.*)

SUPRA-MARGINAL LOBULE.

Dritte Scheitellappenwindung .. .. (*Wagner.*)
Unterzug aus der hintern Centralwindung }
Scheitelhöckerläppchen .. .. .. } (*Huschke.*)
Pli marginal supérieur .. .. .. (*Gratiolet.*)
Erste oder vordere Scheitelbogenwindung .. (*Bischoff.*)

ANGULAR GYRUS (*Huxley*).

Unteres Scheitelläppchen (posterior part) .. (*Ecker.*)
Zweite oder mittlere Scheitellappenwindung (*Wagner.*)
Aufsteigende Windung zum hintern aussern }
Scheitelläppchen und .. .. .. } (*Huschke.*)
Hinteres ausseres Scheitelläppchen . .. }

Zweite oder mittlere Scheitelbogenwindung   (*Bischoff.*)
Pli courbe   ..   ..   (*Gratiolet.*)

QUADRATE LOBULE (*Huxley.*)

Vorzwickel   ..   ..   (*Burdach.*)
Lobule quadrilatère   ..   ..   (*Foville.*)
Præcuneus   ..   ..   (*Burdach.*)

CONVOLUTION OF CORPUS CALLOSUM.

Gyrus Fornicatus—Bogenwulst   ..   ..   (*Arnold.*)
Gyrus cinguli Zwinge or Cingula   ..   ..   (*Burdach.*)
Callosal gyrus ..   ..   (*Huxley.*)
Circonvolution de l'ourlet   ..   ..   ..   (*Foville.*)
Processo enteroidio cristato ..   ..   ..   (*Rolando.*)

UNCINATE GYRUS (*Huxley—Turner*).

Gyrus Hippocampi   ..   ..   (*Burdach.*)
Subiculum Cornu Ammonis ..   ..   ..   (*Burdach.*)
Circonvolution à crochet   ..   .,   ..   (*Vicq d'Azyr.*)
Pli Unciforme ..   .. ⎫
Temporal moyen interne   ..   ..   .. ⎬ (*Gratiolet.*)
Lobule de l'hippocampe   ..   ..   .. ⎭

DENTATE GYRUS.

Corps godronné..   ..   ..   (*Gratiolet, etc.*)

CUNEUS (*Burdach, Turner, etc.*)

Zwickel ..   ..   (*Ecker.*)
Erste oder Hinterhauptlappenwindung   ..   (*Wagner.*)
Oberer Zwischenscheitelbeinlappen ..   ..   (*Huschke.*)
Lobule occipital   ..   ..   (*Gratiolet.*)
Internal occipital lobule   ..   ..   ..   (*Huxley.*)

FIRST ANNECTANT GYRUS.

Erste oder obere Hinterhauptwindung   ..   (*Ecker.*)
Oberer Zug der hintern Centralwindung ..   (*Huschke.*)
Erste obere Hinterlappenwindung ..   ..   (*Wagner.*)
Obere innere Scheitelbogenwindung   ..   (*Bischoff.*)
First external annectant gyrus   ..   ..   (*Huxley.*)
Pli de passage supérieur externe   ..   .. ⎫
Pli occipital supérieur..   ..   ..   .. ⎭ (*Gratiolet.*)

SECOND ANNECTANT GYRUS.

Zweite oder mittlere Hinterhauptwindung..   (*Ecker.*)
Zweite mittlere Hinterlappenwindung   ..   (*Wagner.*)
Pli occipital moyen   ..   .. ⎫
Deuxième pli de passage externe   ..   .. ⎭ (*Gratiolet.*)
Gyrus occipitalis medius   ..   ..   ..   (*Pansch.*)

Medio-occipital and second external annectant   ..   ..   (*Huxley.*)

THIRD AND FOURTH ANNECTANT GYRI.

Dritte oder untere Hinterhauptwindung ..   (*Ecker.*)
Dritte untere Hinterlappenwindung   ..   (*Wagner.*)
Pli occipital inférieur   ..   ..   ..)
Troisième et Quatrième pli de passage   (*Gratiolet.*)
  externe   .. ..)
Gyrus occipitalis inferior .. .. .. (*Pansch.*)

LOBES OF THE CEREBRUM.

Frontal .. .. = Scheitellappen.
Parietal .. .. = Stirnlappen.
Temporo-sphenoidal .. .. .. = Schläfenlappen.[1]
                         Keilbeinlappen.[2]
                         Schläfen-Keilbeinlappen.[3]
Occipital.. .. .. = Hinterhauptlappen.
Central Lobe .. .. = Zwischenlappen.
                         Versteckterlappen.
                         Centrallappen.
                         Stammlappen.
                         Insel.

[1] = Temporal   [2] = Sphenoidal.   [3] = Temporo-sphenoidal.

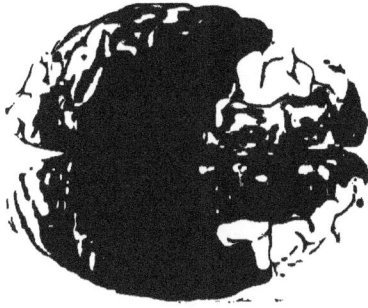

FIG. 4.—CONVOLUTIONS OF THE CEREBRUM, AS SEEN
AT THE VERTEX (*after Bischoff*).

FIG. 5.— CONVOLUTIONS OF THE CEREBRUM, AS SEEN
AT THE MEDIAN ASPECT (*after Bischoff*).

FIG. 6.—CONVOLUTIONS OF THE CEREBRUM, AS SEEN
AT THE BASE *(after Bischoff)*.

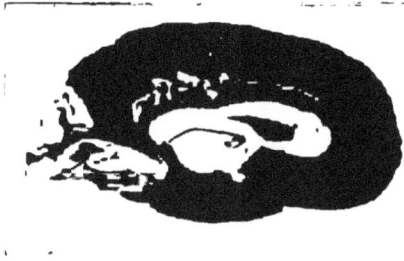

FIG. 7.—CONVOLUTIONS OF THE CEREBRUM, LATERAL
ASPECT (*after Bischoff*).

PART II

MINUTE EXAMINATION OF THE BRAIN

PRIOR to considering in detail the methods adopted for the minute examination of the brain, the student should be fully impressed with the conditions which render many of the processes applicable for other tissues quite inadmissible here. The consistence of cerebral tissue is such that cutting fine sections by the hand, either with an ordinary razor or the Valentin blades, cannot be adopted with any hope of success, since the finest sections thus obtainable are useless for microscopic examination, which requires sections of extreme tenuity. These extremely fine sections can only be obtained by aid of the screw microtome from portions of hardened or frozen brain. In order to modify the consistence of the tissue so as to adapt it for this purpose, various methods of hardening by chromic acid, chromic salts, alcohol, picric acid, and osmic acid have been devised. The results which have accrued from these methods have been most valuable, yet they are uniformly open to the serious objection that this altered consistence induced by the reagent is obtained at the expense of modified structure and altered relationships. Amongst the most serious drawbacks to the use of corrugating reagents in the examination of the brain are the following :—

A considerable expenditure of time (from three to eight weeks being required for satisfactory hardening).

The process itself is extremely tedious, and often unsatisfactory in its results, as evidenced by experienced histologists.

It requires a considerable amount of practised manipulation, and exposes the tyro to numerous disappoint-
ments and failures.

The shrinking of tissue is a most objectionable feature in
the cortex, often amounting to one-half the original
bulk.

The normal wealth of structure is greatly modified; in
this respect it is infinitely surpassed by fresh methods.

The absolute and relative depth of the various layers of the
cortex is subjected to most objectionable modifications.

It affects the cortex of different animals to a different
extent, thus interfering with comparative investigations. Many pathological and normal appearances
are wholly obliterated by the " hardening processes."

Of all tissues, that of the brain, from its extreme delicacy
and susceptibility to rapid post-mortem change, is the tissue
which, above all others, demands a rapid, ready, and fresh
method of preparation, the employment of indifferent media,
and the restriction of all corrugating reagents. No method
as yet adopted can surpass for elegance, expedition, and
certainty the freezing method; and, above all, it is *the*
method which should be chiefly trusted to in all exact
anatomical and pathological research. Yet, whilst we claim
for the fresh freezing methods such self-evident advantages
over the older process, the long series of objections to the
latter tabulated above must not induce us to blind our eyes to
the fact that it also possesses its own advantages and cannot
be dispensed with. Most of our classical descriptions of
cerebral structure refer to brain which has been subjected to
these methods of preparation ; and few indeed, far too few,
are the delineations of the minute structures of the cortex
and medulla in the perfectly fresh brain. Hence, as a
question of simple comparison between the results of different
observers and our own, the hardening methods must still be
adopted. There are numerous other very apparent reasons

why we cannot afford to dispense with the older processes, and the student is therefore recommended, whilst guarding himself from its fallacies, checking or confirming its results by the fresh methods, to make himself equally familiar with the practical details of both methods, and to learn to discriminate the special work to which each process is more particularly suited.

Two Distinct Methods of Freezing.—Just as we find amongst the list of reagents used for hardening brain, some, and especially osmic acid, far less open to the objectionable qualities possessed by alcohol and chromic acid, inducing in fact very slight alteration in the relationships and bulk of the tissue, so do we find the different methods of freezing possess their own distinct and relative merit. This fact, which it is all-important to recognize when dealing with structures like the brain, has been wholly misunderstood or overlooked by all authorities who have written upon the subject, and we constantly find histologists grouping the freezing by ice and salt with the ether freezing method, as though both were equally suitable for all tissues alike. Now the fact is that the ordinary *ice and salt* freezing microtome is entirely useless for brain-structures, except after a modification of the latter by a process of hardening, and hence no longer a *fresh* method. Upon the other hand, the ether freezing microtome is applicable to all tissues having a consistence not above that of the liver, but is pre-eminently adapted for nervous tissue. I find it necessary to insist upon this point, since it seems generally understood that the ice and salt mixture is suitable for freezing brain. Prior to the introduction of freezing by ether, practised histologists constantly complained of the impossibility of obtaining fine sections of brain by freezing, the objections being that hard spiculæ formed within its structure, and tore it up on cutting through it. Very lately this fact has been again asserted by a skilled manipulator,[1] who has even introduced a method of

---

[1] Vide Hamilton in the *Journ. Anatomy and Phys.* vol. xii. p. 259.

combined hardening and freezing to overcome this very unsatisfactory action. Hitherto, therefore, it may be asserted that the use of the ice and salt microtome has failed for the purposes of the cerebral histologist; and although by further development it may be rendered subservient to his purposes, for the present the freezing of brain-substance must be uniformly pursued upon the ether freezing microtome. The methods available for the examination of the minute structure of the brain are as follows:—

Hardening processes for sections to be cut on the imbedding microtome.

Hardening processes for sections to be cut on the freezing microtome.

Fresh process by means of the ether freezing microtome.

Fresh process by a modified teazing and staining.

Fresh process by ordinary teazing with dissociating reagents.

We will preface our account of these methods by a brief outline of the more useful microtomes used.

CHAPTER VIII.

MICROTOMES FOR IMBEDDING.

AMONGST the various instruments devised for cutting sections of hardened tissues are the microtomes of Henson, Rivet-Leiser, Brandt, Roy, His, Ranvier, Stirling, and Rutherford. The instrument more generally used in this country is that devised by Stirling, and its modification by Rutherford, either of which instruments will prove satisfactory to the student; whilst that of Ranvier may be employed where it is desirable to grasp the microtome in the hand, or immerse it during section-cutting in water or spirit.

The principle adopted in Stirling's instrument has been very generally followed. In the last three named, the body of the instrument consists of a metal tube or hollow cylinder, fixed to a table by some simple arrangement, or held in the hand in an upright or vertical position. A fine micrometer-screw works into the lower end, driving through the hollow of the cylinder a closely-fitted piston-plug, which in its turn propels the mass in which the tissue to be cut is imbedded. The essential portion of this form of microtome, therefore, is the body or cylinder, and the powerful micrometer-screw. The body, or cylinder, includes the following divisions:— the hollow, or "well" of the cylinder; its closely-fitted plug, or "piston;" the upper smooth and levelled extremity, or "section-plate;" the lower extremity forming a "female screw" for the reception of the micrometer-screw. It is far preferable that the cylinder should be clamped by projecting arms and screw to a firm support as a table. In choosing a microtome for cutting sections of an imbedded tissue the student must be guided by the following considerations. The instrument should be strong but compact, and not weighty or cumbersome; it should possess appliances for fixing it firmly and immovably to a supporting ledge or table. Its well should be at least one inch in diameter, and the oval section is to be preferred to the circular well, since this form of well wholly prevents rotation of the imbedded mass. The section-plate should be absolutely level, perfectly smooth, polished, and show no irregularities of surface or indentations around the margin of the well, which otherwise would ruin the edge of the blade. The micrometer-screw should be of powerful leverage, work evenly, easily, and without the slightest "loss of time," and have a pitch of at least fifty threads to the inch. Allowing his judgment to be guided by the above rules, the student cannot err greatly in his selection of a microtome for ordinary work and the smaller class of sections. A few words on the different varieties of instruments used may prove of service to the student.

1. **Stirling's Microtome** is a compact, strong, and admirably finished instrument, embracing all the qualities essential for the section-cutting of hardened preparations. For small sections no better adapted form has yet been devised, and for his earliest attempts in cutting hardened brain the student is recommended to secure this form of microtome. This instrument is made to be clamped to a table, so that both hands are free for section-cutting.

2. **Ranvier's Microtome.**—This is a smaller but very useful instrument. It has, however, to be held in the hand, a defect which is in part counterbalanced by the ease with which it can be immersed in spirit, and sections cut whilst so situated. It is well adapted for sections of spinal cord and the large nerve-trunks.

3. **Rutherford's Microtome.**—This, which is a modified form of Stirling's microtome, is a most valuable instrument,

as it is well adapted for section-cutting of imbedded or of frozen tissues. It will be described fully in the section on freezing microtomes. It has the disadvantage of being somewhat cumbersome as compared with the smaller " Stirling," whilst at the same time as a freezing instrument for nervous tissues it is excelled by the ether freezing microtomes. The latter remark does not apply to other animal tissues.

4. **Roy's Microtome**[1] consists of an inner vertical brass plate covered with a layer of cork, and sliding by vertical movement within an outer brass frame-work. A glass rod of horse-shoe form is fixed horizontally by its extremities into the brass frame, upon which the blade glides during section-cutting. The imbedding mixture is first cast in a zinc mould, the tissue being placed in it, and when hardened, the mass is fixed upon the cork plate. The vertical movement of the brass slide is obtained by propulsion from a fine-threaded screw fixed below. A simple tubulature, adapted to the instrument, allows spirit or water to be blown upon the mass when required. It has been stated that tissues may be cut fresh with this microtome after freezing by the direct application of the ether spray. The student must be cautioned against any such attempt; for all tissues alike, the direct application of ether is to be deprecated, and for brain and spinal cord and all nervous structures such a method is wholly inadmissible. Roy's microtome has the disadvantage of not being readily fixed to a table during manipulation.[2]

5. **Schiefferdecker's Microtome** is described here as applicable to hardened preparations, although it is essentially constituted to dispense with the method of imbedding. The object to be cut is fixed within a hollow brass cylinder by means of a clamp upon its surface worked by two screws. In lieu of a propelling screw below, as in other microtomes, a circular plate forming the section-plate can be elevated or lowered by the revolution of an outer concentric plate which

---

[1] Described and figured in the *Journal of Physiology*, vol. ii. No. 1.
[2] The instrument is made by Mr. Gardner, South Bridge, Edinburgh.

turns upon a screw. The student will find this microtome
figured and described in the *Quarterly Journal of Microscopical Science* for January, 1877.

6. Microtome for Slicing through Whole Hemisphere of Human Brain.—The instrument which I have
had made for this purpose at the West Riding Asylum consists
of a heavy brass cylinder, 3 inches deep and 8 inches in
diameter, closed in below, where, however, its cavity communicates centrally with that of a small secondary cylinder—
the screw-socket. In this socket, which is 4 inches deep,
works a powerful and finely-threaded screw, having a milled
head 3 inches in diameter. A flat, circular brass plate,
accurately fitted to the interior of the upper cylinder or well,
is raised or lowered by means of the screw, the movement
being communicated to the imbedding mass, which rests
upon it above. The section-plate is constituted by the
projecting rim of the cylinder above, and is mathematically
level and smoothly polished. The microtome rests upon the
iron collar of a powerful tripod stand. A zinc tray, $2\frac{1}{4}$ inches
deep and 26 by 17 inches, can be adapted to the microtome-
cylinder, so that sections may be cut under water with the
greatest ease.

MICROTOMES FOR FREEZING.

1. The Ether Freezing Microtome (Bevan Lewis).
—This microtome was described in the *Journal of Anatomy
and Physiology* for April, 1877, but has been modified and
improved in several respects by the present maker.[1] It
consists essentially of *a*, the body; *b*, the freezing chamber;
c, the section-plate.

a. The Body.—This is really a modified "Stirling microtome," and forms the lower half of the instrument. Through
its central aperture works a brass plug, driven by a strong
but fine micrometer-screw. In my own instrument, the pitch

[1] The improved microtome may be obtained of Mr. Gardner, 45, South
Bridge, Edinburgh, who, at a minimum of charge, has fully perfected the
workmanship. Descriptions of the microtome may be also found in *Brain*,
October, 1878; and in Dr. Stirling's "Text-book of Practical Histology."

of the screw is fifty threads to the inch; the screw has a diameter of $\frac{8}{10}$-inch, with a milled head, 1½ inches across.   A coarse screw secures the body of the instrument to a table.

b. *The Freezing Chamber.*—This consists of a zinc cylinder, closed above and below, pierced on either side by a large

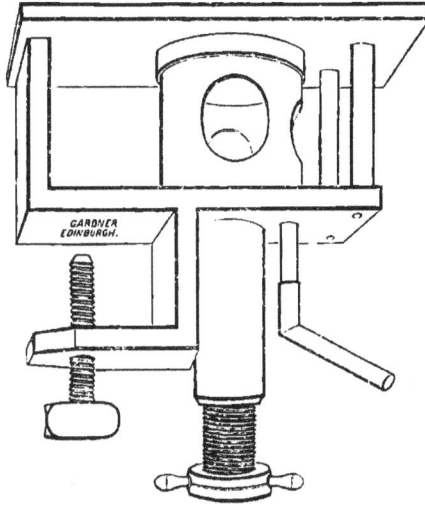

FIG. 8.—THE ETHER FREEZING MICROTOME.   (Original Model.)

aperture, to allow of the admission of the nozzle of the spray-producer, as well as to permit free evaporation of ether, much of which, however, condenses on the bottom, and is conveyed off by the bent tube into a bottle attached to it.   I have found it convenient to have the zinc cylinder quite 2¼ inches wide by 1½ inches deep, and the aperture in the side of the instrument ¾-inch in diameter.

The cap or plate closing the cylinder above, I term the freezing plate, and upon it rests the tissue, whilst the ether spray plays upon the plate from below.   The freezing chamber is thus rapidly reduced in temperature, and to prevent conduction by means of the section-plate, it is well to have a free interval all around the freezing cylinder, securing it from actual contact with the section-plate.

c. The Section-Plate.—This is made either of metal or of
plate-glass, drilled with a central aperture, through which the
freezing chamber glides. If of metal, it should, by preference, be made of a smooth, polished zinc plate, about $\frac{2}{3}$-inch
thick, supported by strong vertical arms upon the body of

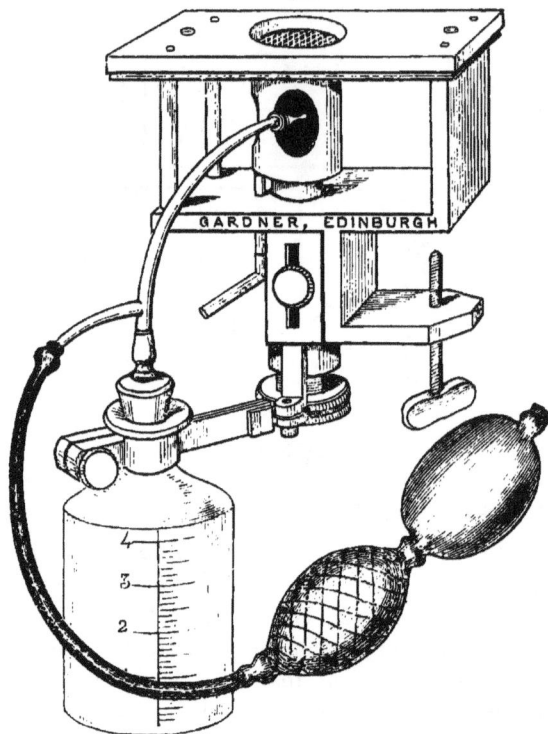

FIG. 9.—THE ETHER FREEZING MICROTOME. (New Form.)

the instrument. I would strongly advocate the use of the
plate-glass, since it renders the movements of the knife absolutely free and easy; whilst, on the other hand, the blade
never gets injured from scratches, which, sooner or later,
invariably appear upon a metal section-plate, and which turn
or indent the edge.

To recapitulate—the various divisions of our microtome are as follows :—

a. The "body," comprising the "well," the "piston," and the "micrometer-screw."

b. The "freezing chamber," fitted with sloping false-bottom, an "exit-tube," and capped by the "freezing plate."

c. The "section-plate," drilled for the passage of the "freezing chamber."

The Freezing Medium.—The best anæsthetic ether should be employed. I find it more economical than methylated ether. The ordinary Richardson ether spray is the more generally used instrument ; but a further improvement has been introduced by the maker, in the form of a lateral support for the ether-spray apparatus, communicating with a bellows worked by the foot. In this form, the spray adjustment is easy, and the hands are both free for section-cutting and manipulation, and thus a most substantial and valuable improvement has been obtained, with but a very trifling addition to the cost of the microtome.

2. **Rutherford's Freezing Microtome.**[1]—This instrument consists of a brass section-plate, with a central aperture leading into the interior of a vertical tube, in which a plug is fitted to move upwards or downwards by means of a fine screw. The cylinder is surrounded on all sides by a metal box covered with gutta-percha, which holds the freezing mixture of ice and salt. An exit tube allows of the escape of water. The whole instrument can be securely clamped to a table. The method of using this instrument for freezing is as follows :—The plug is first unscrewed and oiled, so as to prevent its fixture during the process of freezing. The tissue to be frozen and cut is first immersed in a thick

[1] "A New Freezing Microtome : " *Monthly Micros. Journal*, vol. x. p. 185. "Outlines of Practical Histology : " by William Rutherford. M.D. 2nd edit. p. 164.

solution of gum for some hours, "in order that the gum may permeate every part of the tissue, and prevent the formation of a crystalline condition within the frozen tissue." If the tissue has been previously immersed in alcohol, all traces of the latter should be first removed by soaking in water. Equal parts of *finely-powdered* ice and salt are now placed in the freezing box and stirred around the well, whilst the latter is filled with the solution of gum. As the gum freezes around the periphery, the tissue is plunged into it, and held

FIG. 10.—RUTHERFORD'S FREEZING MICROTOME.

until it is fixed by the advancing ice. Care should be taken to keep the exit-tube of the box open, so as to allow of free escape of water, and to close the freezing box by a weighted strip of cork, to prevent the entrance of heat and to exclude the salt of the freezing mixture. By this method the freezing process may be accomplished in from ten to twenty minutes. All delicate tissues require a *special preparation* to adapt them for this process, otherwise the water which they contain freezes into hard icy spicules, which tear the structure when it is cut. Dr. Pritchard has therefore recommended a prior immersion of the tissue in thick solution of gum-arabic, using the same solution as the imbedding material.

Quite recently, Dr. Hamilton writes as follows with regard to freezing the nervous structures, and especially brain, by means of the ice and salt microtome. " It was found that the crystals of ice so broke up the delicate nervous tissue as to render it totally useless for minute examination. . . . . I attempted two years ago to modify the method of freezing, but without success, and accordingly gave it up as an almost hopeless task." [1]

According to this statement, the writer's experience tallies wholly with my own; and fully acquainted as he was with the methods of steeping tissues in gum prior to freezing which were adopted by Dr. Rutherford, it is clear that this method of freezing was not applicable for fresh brain, although it answers admirably in the case of firmer and less delicate textures. Dr. Hamilton overcame the difficulty of freezing brain with the ice and salt mixture, but at the expense of sacrificing the fresh for the chrome hardening process, a special procedure being requisite, which will be considered further on. Now it is after such considerations that we recognize the undoubted superiority of the ether freezing microtome for nervous structures. It has been stated that the ether freezing process has been entirely superseded by the introduction of ice and salt, or other mixtures, which by *constant* refrigeration keep the mass in a frozen state for hours. All practical manipulators, however, know that this is, as regards nervous structures, a fallacy and a blunder, since the very excellence of the ether freezing method depends upon the fact that it can be checked at any stage and renewed when required. This is the all-important consideration in freezing brain, for beyond certain limits it is frozen into a hard icy solid, which at once blunts or turns the edge of the blade; but with the ether-spray this stage need never be attained, and a consistence is obtained admirably adapted for

[1] *Journ. of Anat. and Phys.* vol. xii. p. 257.

section-cutting. The student should therefore bear in mind
that *the constant application of a freezing mixture to fresh brain
cannot but result in failure for section-cutting;* and that *by use
of the ether process the prior preparation of brain by hardening,
or by immersion and saturation in mucilaginous fluids, may be
wholly dispensed with.*

3. **Williams' Freezing Microtome.**[1]—In this instrument a wooden tub containing a freezing mixture of ice and

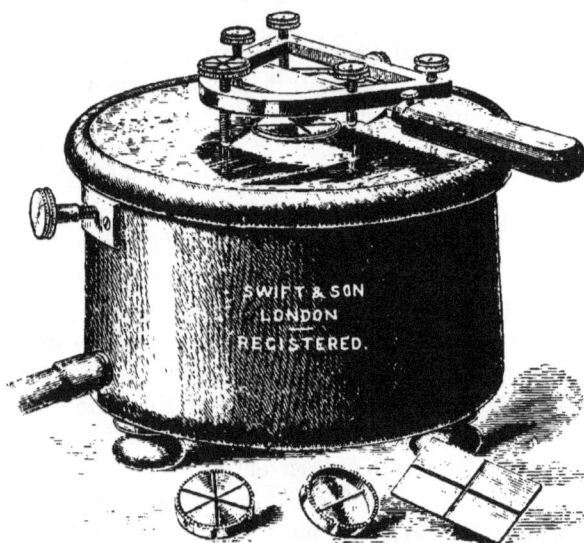

Fig. 11.—WILLIAMS' FREEZING MICROTOME.

salt is covered by a glass lid, the frame of which is secured
by a clamp screw. An upright brass conducting-bar passes
from the interior of this box through a central aperture in
the lid, and to the extremity of this bar is screwed the
circular brass plate which supports the tissue to be frozen.
The arrangement of the knife is peculiar, since it is fixed in

[1] This instrument is made by Messrs. Swift & Son, University Street,
London.

a triangular frame, and can be either raised or lowered by the screws which are adapted to the frame. There are three additional plates for supporting the frozen object, and a brass cup for holding substances, which are fixed in cacao-butter or paraffin.

The box is filled with equal parts of pulverized ice and salt, care being taken to prevent the mixture touching and so fixing the cover. After the cover is replaced and screwed down, the substance to be cut is placed on the central brass plate, surrounded by a little solution of gum, and the apparatus is covered with baize to facilitate freezing. When frozen, raise the blade, and after the first cut across, proceed as follows :—Each end of the razor must be presented to the surface of the section, and *exactly levelled by means of each of the back screws.* If the large back screw be now turned, the blade can be lowered to any required extent ; and since a complete revolution of the screw gives us a section $\frac{1}{100}$-inch in thickness, and the screw-head is graduated into sixths—a movement through sixty degrees gives us a section $\frac{1}{600}$-inch thick.

The special features of this microtome, in which it differs materially from others already described, consist in an arrangement whereby the blade and not the imbedded or frozen tissue becomes the movable part of the instrument ; and the edge of the knife is only brought into contact with the substance to be cut.

H 2

CHAPTER IX.

IN subjecting the brain to the agency of hardening reagents certain important conditions should always be kept in mind, as they are essential to success. The conditions are as follows :—

a. The reagent should act equably upon all portions of the tissue.

b. The requisite consistence must be acquired at the expense of the minimum of alteration and shrinking of tissue.

Now the first of these conditions can only be obtained by ensuring a thorough saturation of the tissue throughout, so that the fluid permeates rapidly to the central or deepest portions of the mass. It is evident that the surface of the tissue being bathed in the reagent will be more actively affected by the latter than the more distant parts within, and thus arises the danger of a too rapid hardening of the exterior, which forms a mechanical impediment to the permeation of the deeper structures by the surrounding fluid. Always, therefore, take the precaution to ensure a *free and rapid permeation of the tissues by the fluid.*

Again, if any portion of the surface is in close contact with the side of the containing vessel, it necessarily is less affected than the surface bathed in the fluid. The mass should therefore be so suspended that it is *equably bathed all around in the reagent.*

Another important consideration is that of temperature, for under the most favourable circumstances the central or deeper structures must remain far less subject to the action of the reagent than the exterior of the mass, and are, therefore, the parts which most readily succumb to putrefactive changes. The larger the mass, therefore, to be hardened, the more difficult is it to prevent central decay, whilst an elevated temperature induces the same result. The tissue to be hardened should therefore be of *moderate bulk* as compared with the fluid in which it is immersed, and the preparation should be kept in a cool spot, or better still, *in an ice safe*.

The next condition we have named, and which it is equally important to secure, is that a minimum of shrinking of tissue should result from the action of the reagent, and it is a well-known fact that all the reagents used for this purpose will, if employed in too concentrated a form, ruin the preparation by inducing extreme shrinking and brittleness of the mass; whilst again, some of these reagents are far more reliable than others, and less open to these disadvantages. The more commonly used hardening reagents are—osmic acid, Müller's fluid, solutions of the chrome salts, chromic acid, picric acid, methylated spirits, and alcohol. Now this list represents their relative value as hardening reagents for nervous tissues, osmic acid and Müller's fluid standing at the head of the series as the most valuable and least injurious in their action, chromic acid and alcohol occupying a far less prominent position in the scale. In order therefore to secure the tissue from injurious shrinking, employ in preference to the others the *reagents noted at the head of the list,* and use the *weakest solutions compatible with safety to the tissue,* commencing with the weakest, and gradually augmenting the strength of the solution, or, later on, even replacing by the more astringent reagents. Let us now detail throughout the process recommended to the student for his first essay in chrome-hardening, subsequent to which it will be useful to briefly dwell upon the various modifications of the process adopted by others.

§ PROCESS OF HARDENING BY CHROME.

1. Müller's Fluid and Potassium Bichromate.— Excise a portion of the ascending frontal or parietal convolution of human brain, cutting across its length so as to remove about an inch of the convolution along with its central and deeper medulla. Lightly cover it all round with a little cotton-wool, and immerse it in from two to three ounces of methylated spirits contained in a four-ounce stoppered bottle.

Label the bottle with the name of the specimen and date, placing it in a cool cellar or ice safe.

In twenty-four hours pour away the spirit, replacing it by four ounces of Müller's fluid, the preparation being surrounded as before by cotton-wool to ensure it being bathed upon all sides alike by the fluid.

Let the bottle stand in a cool spot, and in three days replace the fluid by a fresh quantity. At the end of one week the fluid should be again renewed, or, preferably, a weak solution of potassium bichromate substituted (2 per cent.). At the end of the second week a solution of the latter of double the strength may be added; and if at the termination of the third week the mass is still pliable, and of the consistence of ordinary rubber, it is as yet unfit for section-cutting, and the reagent should be replaced by a solution of chromic acid.

In these later stages the chromic acid expedites the process without producing the extreme shrinking of tissue which ensues if it be used at earlier stages. For the same reason even absolute alcohol may at this stage be employed, but although valuable in the preparation of the spinal cord, it cannot be equally well recommended for the cortex of the hemispheres. By the above process our preparation will have attained the requisite consistence within a period of from four to eight weeks. It will be observed that in the various steps of this process the conditions previously emphasized as essential to success are obtained. Thus the early immersion in methy-

lated spirit abstracts all superfluous fluid from the brain and
its vessels, and entering its substance by its affinity for water,
aids in the rapid permeation of the mass by the chromic solution in the next stage. The more prolonged action of the
spirit would, however, prove highly detrimental. Again, the
cotton-wool ensures an equable distribution of fluid around,
whilst the comparative bulk of the mass and fluid in which it
is immersed, and the temperature to which it is exposed,
provide for equable hardening to its deepest structures, and
ensure it against decomposition. The gradual increase in the
potency of our reagents from Müller's fluid to the strong
solutions of chrome is also an adjunct to the hardening of
the central portions, whilst we also expedite the process
of hardening.

2. **Potassium Bichromate and Chromic Acid**
(Rutherford).—Place small portions of the cerebrum in
methylated spirit for twenty-four hours, observing the same
precautions as to relative bulk of the preparation and the
reagent, covering with cotton-wool, and leaving in a cool
place. Replace the spirit by a mixture of potassium bichromate and chromic acid. The proportions may advantageously
be varied, according to the condition of the structure to be
hardened; but the solution recommended by Rutherford, and
which answers well, contains 1 gramme of chromic acid and
2 grammes of potassium bichromate to 1,200 c.c. of water.
Change at the end of eighteen hours, and then once a week.
Should the tissue not be sufficiently tough for cutting at the
end of six weeks, place it in a ½ per cent. solution of chromic
acid for a fortnight, and then in rectified spirit.

3. **Iodized Spirit and Potassium Bichromate** (Betz).
—Large portions of cerebrum may also be placed for a few
hours in methylated spirit, tinted of a light sherry-brown by
tincture of iodine. Add fresh iodine solution as the colour
fades. In one or two days remove the pia mater, and return
the preparation to the solution, adding to the latter half its
bulk of fresh iodized spirit. After the lapse of another period

of two days, replace the solution by iodized alcohol (alcohol
70°—80°) tinted of a sherry-brown by a tincture of iodine.
In from two to three days it should be transferred to a 4 per
cent. solution of potassium bichromate until sufficiently
hardened for section-cutting. Should a brown deposit form
over its surface during the process, let it be well washed, and
a fresh solution used. When hardened, these preparations
may be kept permanently in 0·5 per cent. solution of bichromate. If the cerebellum is to be hardened, the segments are
at once placed in the iodized alcohol, adding fresh iodine
frequently as the colour of the solution pales. Remove the
pia mater on the second or third day, and in a week transfer
to pure methylated spirit for twenty-four hours, and finally
harden it in a ·5 per cent. solution of potassium bichromate.
This process, recommended by Prof. Betz, is not suitable
for examination of the cortex, as the iodized spirit is injurious
to the after-processes of staining. It is, however, especially
suited for obtaining large sections through the hemisphere,
the whole cerebrum and cerebellum, when sliced evenly across
into segments ¾-inch thick, being most satisfactorily hardened
throughout. For the minute examination of the cortex
methods 1 and 2 are pre-eminently to be preferred.

4. **Müller's Fluid and Ammonium Bichromate**
(Hamilton).—This method is especially applicable to large
segments of the brain, and is much to be preferred to the
process by chromic acid, 5. The brain is sliced completely
through into segments about one inch thick. Each segment
may now be placed in a large vessel, such as a brain-preparation jar, padded with cotton-wool, and containing a comparatively large bulk of the solution, which consists of three
parts of Müller's fluid to one part of methylated spirit. A
refrigerator or ice safe should invariably be employed to pre-
clude decomposition, and the pieces should be turned over
occasionally in the solution. In about three weeks they
may be transferred to a solution of ammonium bichromate
(·25 per cent.). At the end of the fourth week replace by a

1 per cent. solution of the same salt, and the following week by a 2 per cent. solution, in which they remain until fit for section-cutting.

5. **Chromic Acid Solution** (Lockhart Clarke).—The convolutions of the cerebrum and cerebellum were by this process hardened in a 0·25 per cent. solution of the crystallized chromic acid—a stronger solution rendering them far too brittle for section-cutting. As stated above, chromic acid is not adapted for hardening so satisfactorily and uniformly as the chrome salts. This method, employed by Lockhart Clarke, has been superseded by more reliable methods, as are also the processes recommended by Stilling, Kolliker, Hanover, and Van-der-Kolk, for the hardening of nervous textures.

§ HARDENING BY OSMIC ACID.

It has been already stated that osmic acid (OsO_4) is one of the most reliable agents for hardening the brain and other nervous textures which we possess. Prof. Sigm. Exner, of Vienna, has therefore devised a method whereby *small* portions of the brain may be prepared for section-cutting by means of this reagent.[1]

Exner's Process.—A small portion of brain, not exceeding one cubic centimetre in size, is placed in ten times its volume of a solution of osmic acid (1 per cent.). The solution should be replaced by a fresh reagent of the same strength after the lapse of two days—a proceeding which may be advantageously repeated at the end of the fourth day. In from five to ten days the piece is usually stained throughout, for this reagent has the valuable property of hardening and staining simultaneously. The hardened brain is then washed in water, plunged for a second in alcohol to facilitate the imbedding, and sections are cut in the usual way in an

[1] "Zur Kenutniss vom feineren Baue der Grosshirnrinde." Aus dem lxxxiii. Bande der Sitzb. der k. Akad. der Wissensch. iii. Abth. 1881.

ordinary microtome. The subsequent treatment of these
sections will be given later on, when methods of staining and
mounting are considered. It is only necessary here to remind
the student that osmic acid is really a most valuable agent in
the investigation of brain-structure, and he should avail himself of every opportunity of becoming familiar with its action.

§ SUMMARY OF PROCESSES FOR HARDENING.

The student having been thus placed in possession of the
more valuable methods for preparing the brain for the
imbedding microtome, it will be well to indicate the method
he should adopt in his earlier attempts, as each process has its
own individual merits. It is advisable that his first attempts
should be made with comparatively small portions of tissue
—about 3 c.c. in bulk—and that he should employ the
process of hardening by Müller's fluid and potassium bichromate.

1. This process, to which I assign the first place amongst
the chrome-hardening processes, is slow but very certain. The
results are, to my mind, more satisfactory than those of any
of the other methods, and I invariably adopt it myself as the
one for general use at the West Riding Asylum. The vessels
which the student will find most convenient to use for hardening these smaller portions of tissues are the stoppered glass
bottles of 4-oz. capacity employed for dispensing purposes.
Each bottle should be labelled with the name of specimen and
date and nature of each successive change of reagent, whilst
the specimen should occasionally be removed, and its consistence noted so as to familiarize the touch with the increasing firmness of the tissue, and the degree of hardness
requisite. The method recommended by Rutherford (2) is
also very reliable for moderate-size specimens. The methods
advocated by Betz and Hamilton (3, 4) are peculiarly well
adapted for large segments of the brain, a whole hemisphere

being thus readily hardened throughout. When these large masses of tissue have to be dealt with, special precautions are requisite. Thus the hemisphere should be sliced horizontally or vertically into segments not over ¾-inch in thickness. Each segment should rest upon a bed of cotton-wool in the vessel for its reception ; and if two segments are included, another stratum of cotton-wool should be interposed betwixt them. The vessels containing the specimen should be of large size, varying in capacity with the bulk of the latter. Thus a large preparation glass, such as is used for preserving brain in spirit on museum shelves, will be found well adapted for hardening the whole hemisphere ; whilst pickle-bottles, especially those provided with the " patent lever stopper," are most suitable for the basal ganglia, cerebellum, smaller portions of the cerebrum and the pons. Still smaller segments, such as the student will have more frequently to deal with, are best hardened in the 4-oz. stoppered bottle, a good supply of which should be kept on hand. All the foregoing processes are contrasted with that of Exner's (osmic acid) in being most adapted for the demonstration of the nerve-cells and plexuses of the grey matter of the brain and its ganglia ; whilst the latter, failing in this respect, is, on the other hand, infinitely better suited for exhibiting the structure of the cerebral medulla and its extensions into the cortex of the brain. The student is therefore recommended to make himself acquainted with the structures of the cortex by the chrome methods, and subsequently to employ Exner's method for demonstrating the medullated tracts of the cortex.

§ IMBEDDING AND SECTION-CUTTING.

Having satisfied himself that the specimen is sufficiently hard for section-cutting, the student first cuts off a portion which can be accommodated by the well of the microtome, and prepares it for imbedding by a prior immersion for a few minutes in rectified spirits. This removes the fluid derived from the chrome solution from which it has been taken. In the meantime the imbedding mass should be melted at the lowest temperature requisite. The micrometer-screw should be lowered sufficiently to allow of a deep imbedding of the tissue. The preparation is next removed from the spirit, rapidly dried by a fold of blotting-paper, and momentarily plunged into the warm imbedding mixture, which, upon its removal, leaves a film over its surface, whilst all small cavities or lacuna which it may contain are filled up. The melted mass is now poured into the well of the microtome, and the preparation immersed and held in the position required for section-cutting until fixed by the cooled and consolidated mass. If the imbedding mass is one subject to much contraction on cooling, it is requisite, just as it is becoming " set," to press down with the end of a spatula the margin of the wax and the oil mass against the sides of the well.

The Imbedding Mass.—In the process of imbedding, the principle to be remembered is that the mass should be not too resistant to the blade, whilst at the same time it affords efficient support for the imbedded tissue. If the texture be one readily permeated by the imbedding mixture, contraction of the latter in cooling is apt to result in injury to the preparation, more especially if this be one of the more delicate structures. Thus, in the delicate embryos of the fowl and other similar organisms, special methods of imbedding will be requisite. For the brain and spinal cord, however, the process is comparatively simple, since the imbedding mass neither permeates its structures nor injures it by contracting.

The following are the more important media employed for imbedding prior to section-cutting :—

1. White wax.   Olive oil, equal parts.
2. White wax, 3 parts.   Olive oil, 1 part.
3. White wax.   Cacao-butter.
4. Solid paraffin, 5 parts.   Hog's lard, 1 part.
5. Solid paraffin, 5 parts.   Paraffin oil, 1 part.   Hog's lard, 1 part.
6. Solid paraffin, 5 parts.   Spermaceti, 2 parts.   Hog's lard, 1 part.
7. Spermaceti, 4 parts.   Cacao-butter, 1 part.
8. Spermaceti, 4 parts.   Castor oil, 1 part.

The first on this list is perhaps the more generally employed medium, but it possesses the disadvantage of contracting strongly on cooling, so as to leave an interval betwixt the wax plug and the sides of the well : this, of course, will necessarily tend to loosen the mass, and allow rotatory or " wobbling " movements, which prevent accurate section-cutting. These disadvantages may be overcome by one or other of the following measures :—

> Vary the proportion of the oil and wax in favour of the former. When solidifying, press down the edge of the mass with a spatula. Use a microtome plug with groove or projections on upper surface. Drive the mass out of the well, partly surround it with a narrow strip of blotting-paper, and return it—forcibly pressing it down into the well—the blotting-paper swells by imbibition of the spirit in section-cutting, and so fixes the plug firmly.

In the use of the third medium, viz., white wax and cacao-butter, far less contraction ensues upon solidification, and a firm supporting mass is obtained, the proportion of the in-gredients varying with the firmness required. I can also speak favourably of the combination of paraffin with hog's lard. In all these cases it is well to keep the mass in a small tinned pot with lip, provided with cover and handle, so that it can readily be held over a gas-jet or Bunsen burner, and melted as required. Another imbedding agent used is gum, which is solidified by the application of alcohol or methylated spirits. This method, proposed by Brücke, although valuable for many tissues, cannot be recommended for brain.[1]

Imbedding is wholly dispensed with in the form of micro-tome devised by Schiefferdecker, where the preparation having been hardened by alcohol as far as practicable, is placed in the well and clamped firmly by the arrangement already de-scribed. In the small hand-microtome of Ranvier, again, the

---

[1] " Handbook of the Physiological Laboratory," p. 92.

ordinary melted media for imbedding are often exchanged
for elder-pith, which is packed around the tissue in the well
in a dry state, and then, upon immersion in spirit, the pith
swells and firmly fixes the preparation. None of these latter
methods are so suitable for hardened brain as imbedding in
the wax and paraffin mixtures.

Section-Cutting.—The preparation being satisfactorily
imbedded, and the mass perfectly cold and hard, our next
procedure is to cut the *finest possible sections—they cannot
be cut too fine.* Place on the table supporting the microtome,
and in front of the latter, two vessels, one a Griffin beaker-
glass of three inches diameter, and the other a cylindrical
glass jar, a brain-preparation jar, or better still, a flat-bottomed
porcelain evaporating basin. Nearly fill the beaker with
methylated spirit, and the latter with water. Seat yourself
in front of the microtome at a convenient level for the free
play of the hands, place a soft towel over the knees, to be
used for wiping off adherent wax from the imbedding mass.
Have close at hand, also, a razor-strop, a porcelain capsule of
small size, and some camel-hair brushes.

Dip the section-blade in the methylated spirit, and having
raised the imbedded mass slightly above the level of the section-plate, slice off the superficial portion, exposing a clean
smooth surface of the tissue for cutting. Again dip the blade
into the spirit, and turning the micrometer-screw, say through
one-eighth of a revolution, place the flat of the blade upon the
section-plate in front of the mass, with the back towards
yourself and the handle in your right hand, grasped near the
heel of the blade. Now cut your first section by a clean
steady sweep from heel to point, and always *away* from yourself. No saw-like movement should be performed, nor should
the blade be arrested occasionally in its course, the section
shaved continuously and evenly off the preparation. The
thin film of wax removed from the front of the tissue is
sometimes likely to get in the way of very fine sections, and
hence it is usually advisable, with a penknife to gently

remove all the imbedding mass *in front* of the preparation
to about the depth of a couple of millimeters prior to section-
cutting. Subsequent sections should be cut with a still more
restricted movement of the screw, until the operator has found
the minimum degree of movement at which a section can be
cut. This, of course, is a matter of tact, and the student
will find that he daily improves the quality of his sections
by practical manipulation. Two precautions must ever be
taken if successful sections are desired. The first is, that as
the blade is raised out of the spirit, the latter should not be
drained off, but a good flow be retained over the surface of
the blade, so that the section is floated up as it is cut. In
the next place make it a rule to pass the blade across the
strop frequently, after cutting half-a-dozen sections, as it
requires to be kept at the finest possible edge. As the
sections are cut they are floated off into the porcelain basin
of water, and are freed from adherent particles of the
imbedding mass by the rapid gyratory movements they here
undergo from the currents caused by the attraction of water
for the spirit they contain. After floating off each section
in the water, pass the blade softly over the towel to clear its
surface from adherent wax, prior to bathing it afresh in the
spirit. How is the student to ascertain whether his sections
be sufficiently thin ? The exact thickness of the section is
ascertained if the pitch of the micrometer-screw is known, and
the degree of movement which he adopts noted. Thus, if the
screw have eighty threads to the inch, a movement through
one-eighth of a complete turn, *i.e.*, 45°, will give a section
$\frac{1}{640}$-inch in thickness, whilst a movement through an
arc of 30°, or one-twelfth the entire turn, gives a section
$\frac{1}{960}$-inch thick. He will, however, be not far from wrong
by following this rule : reject all sections which do not float,
and of the latter preserve as valuable only those which appear
as a mere film on the surface of the fluid, requiring to be
looked at sideways to be distinctly seen. It is at this stage
of the proceeding, that of section-cutting, that we can best
judge of the merits of the imbedding mass, which, if found

unnecessarily resistant to the blade, should be modified by
the addition of more of the diluting adjuncts, viz., lard, oil,
or cacao-butter.

The Section Blade.—This is an item of supreme importance to the histologist, and it is by no means true, as we
often find asserted, that any razor blade answers the purpose
equally well with the specially prepared knives.   I have
found this statement made frequently by those who were by
no means fortunate in the thinness and delicacy of their
sections ; whilst all with whom I am acquainted, who can
rejoice in the good qualities of their sections, have always
paid great attention to the section-blade which they employ.
The qualities of a good blade for section-cutting upon the
microtome depend upon its make, form, edge, and tempering.
It should be sufficiently long to ensure a free sweep over the
whole surface of the section-plate from heel to point, wide
enough from back to edge to support and float up the
largest-sized section which might have to be cut ; it should
be hollowed out on both surfaces, but be most concave on the
surface uppermost in section-cutting ; the edge should be
extremely sharp, and perfect, whilst the angle formed by the
converging surfaces here should not be so acute as to involve
bending of the blade when subjected to slight pressure ; the
quality and tempering of the steel should be of the best
character.   The above points are all most essential require-
ments in a good section-blade ; but another important consideration is that the plane of the back and edge of the knife
should exactly coincide when it rests upon a perfectly level
surface as the microtome section-plate—not the slightest
degree of tilting being admissible.

For the smaller microtomes I have invariably used razor-
blades, my three instruments measuring in length and width
of cutting-surface as follows :—

> No. 1.   4 inches × 1 inch (made by Weedon).
>   4½ ,,   × 1 ,,   (,,   Young).
>   5   ,,   × 1¼ ,,   (,,   Young).

The first is a most valuable knife. It is fitted into a short, firm, and fixed handle. The latter, which is my favourite blade, has a folding handle, and is sketched in outline with a section of its concave surfaces in an early number of *Brain*.[1] Some of the earliest and most discouraging failures upon the part of the tyro in section-cutting depend upon the condition of the blade, and it is necessary that he should be fully impressed with the fact that not only must he secure a couple of section-knives with the above qualities, but that he should jealously keep them in most perfect order. To ensure the latter condition, let him observe the following precautions :—

Use no undue pressure against the section-plate whilst cutting, and keep the surface of the blade absolutely level with the plate. Pass it frequently along the strop whilst cutting sections. After each day's work examine its edge most critically, and if the slightest irregularity or notching be apparent, it must be reset before use.

Keep the blade dry and well polished by a soft handkerchief, and if in a closed handle, have a chamois-leather case made for it.

Subsequent Treatment of Sections.—The sections have now to be removed carefully by a camel-hair brush to methylated spirit contained in the small porcelain capsule, in which they may be allowed to soak for some time to remove all traces of chrome, which otherwise interferes with the subsequent staining they have to undergo.

Large Sections through Hemisphere.—The above account refers especially to the smaller class of sections. When it is requisite to obtain sections upon a much larger scale, such as those through the hemisphere or the whole brain, the process requires modification. The method of imbedding is precisely similar in all its details, but when we come to

---

[1] *Vide Brain*, part iii. p. 353. October, 1878.

I

cutting the section, special forms of knife are found requisite. The blade which I use measures 16 inches long by 2 inches wide, and is fixed in upright handles at each end. The zinc basin in which the large microtome rests is filled with water, and then sections are cut with a saw-like movement of the blade, and not, as in the former cases, by one complete sweep through the structures. It requires a steady hand and much manipulative skill to obtain these larger sections both thin and perfect, and the student will only perfect himself by frequent and persevering efforts. These large sections can be readily removed by means of sheets of paper upon which they are floated, and this support may be given to them throughout the various subsequent stages of preparation. When the desired number of sections have been obtained, the imbedded plug may be removed, the remaining portion of tissue placed in a 0·5 per cent. solution of bichromate of potash, until required. The microtome should be carefully dried and placed away in a drawer for safety, as the slightest scratching of the polished section-plate will ruin a good blade in a few minutes.

Apparatus for Hardening Tissues and Section-cutting.

Winchester bottles for hardening reagents.
Jars and 4-oz. phials for hardening tissues.
An ice safe.
Microtome for imbedding.
Microtome for freezing.
Glass "preparation-jar" for receiving sections.
Imbedding mass.
Spirit lamp.
Porcelain capsule.
Glass beaker for methylated spirit.
Ether spray apparatus.
Section-knife.

§ STAINING AND MOUNTING SECTIONS OF HARDENED BRAIN AND CORD.

The staining reagents advocated by different authorities for sections of the nervous system prepared by hardening are numerous, and the list has become so lengthened since the earlier labours of Lockhart Clarke, Van-der-Kolk, and others, that one great difficulty presented to the student is to make a judicious selection of such as will yield him the best results in the special direction pursued by his work. Amongst those more generally used are carmine and its combinations, picro-carmine, indigo-carmine, and borax-carmine; the aniline series, comprising aniline blue-black, aniline blue, methyl-aniline, rosanilin or magenta; hæmatoxylin or logwood; eosin; picric acid; osmic acid; double chloride of gold and potassium. Many years' experience has convinced me that a large proportion of these reagents may be safely dispensed with. The following really comprises all essential staining solutions for nervous tissues:—

> Hæmatoxylin.
> Carmine in ammoniacal solution.
> Picro-carmine.
> Aniline blue-black.
> Aniline blue.
> Osmic acid.

The only solution comprising a double pigment in this list is the picro-carmine, but double staining may be extensively employed by the combined agency of the above and two other dyes, the more useful being as follows:—

LIST OF PIGMENTS FOR DOUBLE STAINING.

> Aniline picro-carminate.
> Picro-anilin.
> Osmium with picro-carmine.
> Hæmatoxylin with anilin.
> Hæmatoxylin with eosin.
> Picro-carmine with iodine green.

In making choice of metallic impregnation and reduction by osmium or gold, or of staining by the mineral or vegetable pigments in the list given, the student must be influenced by several considerations.

Does he require a uniform staining by a simple dye, or does he wish for differentiation of elements by the use of double pigments—the process of double staining? Does he wish to examine more specially the cortex or medulla of the cerebrum, or the cortex of the cerebellum, or again, the pons, medulla oblongata, or spinal cord?

Does he desire to display certain particular constituents of the cortex—e.g., the nerve-cells and their processes—to the fullest possible extent, even at the sacrifice of other structures, such as is often required in minute investigations into the histological constitution of tissues?

Whatever be his object, these questions must receive an answer before the staining reagent can be selected which will yield him the effect desired. I propose to give here the composition of each reagent, the methods of staining, the special value of each dye for the various histological elements, and to supplement these observations by an analysis of their relative merit for different regions of the cerebro-spinal system.

HÆMATOXYLIN STAINING.

1. **The Dye.**—This reagent has often fallen into great disrepute, owing to the notable variability in the quality of the solutions issued by different makers. The student is strongly recommended to make his own solution, adopting the formula given by Minot or that employed by Kleinenberg. For the brain and spinal cord I use the former, as preferable to that of Kleinenberg, and I find it at all times most trustworthy in its reaction, and in the uniformity of results obtained.

MINOT'S FORMULA.

Hæmatoxylin (crystals)	... ...	3·5 parts.
Absolute alcohol	... ...	100·0 ,,
Alum ...	... ...	1·0 ,,
Water ...	... ...	300·0 ,,

First dissolve the hæmatoxylin in the alcohol, and add to it the alum, previously dissolved in the water. Keep in a closely-stoppered glass bottle labelled "hæmatoxylin dye (Minot)." For use, a solution of alum (0·5 per cent.) is poured into a watch-glass or porcelain capsule, according to the size and number of sections to be stained. A little of the dye is then dropped in, until the solution assumes a light violet tint, when it should be carefully filtered before the sections are immersed. The degree of dilution will soon be learnt after a little practice.

KLEINENBERG'S FORMULA.

a. Make a saturated solution of crystallized calcium chloride in alcohol (70 per cent.), and add alum to saturation.
b. Make a saturated solution of alum in alcohol (70 per cent.), and add Solution 1 to Solution 2 in the proportion of 1 to 8.
c. To the resultant mixture now add a *few* drops of a *barely* alkaline saturated solution of hæmatoxylin.

Excellent as this logwood solution is for embryonic tissues, I cannot recommend it for the sections of hardened brain, such as we are now engaged with, to the same extent that I can Minot's solution.

2. **The Staining Process.**—Withdraw the sections to be stained from the methylated spirits in which they float by means of a glass rod or camel-hair pencil, and immerse them in the logwood dye contained in a watch-glass or porcelain capsule. The sections should be lightly stained, otherwise they are spoilt by too diffuse a colouring and by

acquiring a brittleness unfavourable to subsequent manipulation. In his first attempts the student should examine a section occasionally on the stage of the microscope, and so learn to judge of the progress made by the staining. When sufficiently stained, the dye should be poured off, the sections floated up in a porcelain capsule half full of water, the capsule slightly inclined towards a gentle stream of water from a tap which, falling upon the edge of the vessel, keeps up a constant change of fluid and ensures a most thorough washing of the sections in the currents it produces. Care must be taken that the delicate sections are not torn by this means, which is really necessary to ensure removal of the deposit which adheres to their surface. Next remove the sections to a capsule containing methylated spirit, and in one or two hours all tendency to diffuse staining will have disappeared. They are now placed for five or ten minutes in rectified spirits or absolute alcohol for dehydration, transferred to the centre of a glass slide, floated up by a drop or two of oil of cloves, which, when it has permeated the tissue and rendered it perfectly transparent, is drained off, and the section mounted in Canada balsam.

To ensure successful staining with this reagent, the following points must be carefully attended to :—

Stain only the very finest sections.

Let the sections be perfectly freed by spirit from any acidity (chromic or picric acid, etc.), otherwise the staining will be a failure.

Make your own solution of hæmatoxylin.

Employ Minot's solution in preference to others, and reserve Kleinenberg's for embryonic brain.

Never omit the proper dilution and filtering prior to use.

Guard against diffuse staining and brittleness from too prolonged action of the dye.

Wash the sections *very thoroughly* subsequent to staining.

Remove any accidental diffuse coloration by long immersion in spirit.

Dehydrate perfectly before clearing up with clove oil.
Mount in a benzole solution of balsam.

3. Notes on the Reaction of the Dye.—Hæmatoxylin is specially well suited for the display of the various nuclei met with in brain and spinal cord,—*e.g.*, the connective nuclei of neuroglia, the peri-cellular and perivascular nuclei, the nuclei and nucleoli of the nerve-cells. The larger nerve-cells of the cortex are beautifully shown in successful preparations, and their primary and secondary branches may be followed for some distance, but never to the extent seen in aniline staining. On the other hand, both neuroglia basis and nerve-structures undergo much shrinking by the use of the dye, and from this cause the smaller cells of the upper layers lose, to a great extent, their normal features. The protoplasmic extensions and nerve-fibre network of the cortex are indeed very poorly exhibited in logwood, as compared with aniline preparations. The axial cylinders of medullated fibres exposed in transverse section are well exhibited by logwood, and hence the sections of spinal cord and medulla, stained by the dye, show to advantage.

CARMINE STAINING.

1. The Dye.—The most useful solution for hardened brain sections is that recommended by Beale, diluted to the required strength, the original solution being too strong for our purpose.

BEALE'S FORMULA.

Carmine (in small fragments)	...	10 grains.
Strong solution of ammonia	...	½ drachm.
Price's glycerine	...	2 ounces.
Alcohol		½ ,,
Distilled water ...		2 ,,

Agitate the carmine with the liquor ammoniæ in a test tube gently heated over a spirit lamp. When solution is complete, carefully boil for a few minutes and expose the open test tube for an hour to allow excess of ammonia to escape. Add the water, filter, and after the addition of the spirit and glycerine, expose to the air until the odour of ammonia given off from it is very faint. Keep the clear fluid in a stoppered bottle, and whenever carmine becomes precipitated, add a drop or two of liquor ammoniæ.

FORMULA FOR BORAX CARMINE.

Carmine	1 part.
Borax	4 „
Water	56 „
Alcohol	q.s.

Dissolve the borax in the water, and add the carmine. Filter the solution, and add just sufficient alcohol to ensure free permeation of the tissue by the dye.

THIERSCH'S FORMULA.

Solution A.	Carmine	1 part.
	Liquor ammoniæ	1 „
	Distilled water	3 „
	Dissolve and filter.	
Solution B.	Oxalic acid.	1 part.
	Distilled water	22 „

Mix one part of Solution A with eight parts of Solution B, and add twelve parts of absolute alcohol. Modifications of Gerlach's original carmine dye have also been introduced by Frey and others, but the three above given are all that are really requisite for staining nervous tissues.

2. **The Staining Process.**—Give the preference to Beale's solution, although its density from the glycerine renders the action slow. Dilute the strong solution with seven times its bulk of water and filter. The sections should be placed in a comparatively *large quantity* of the dilute

solution covered from dust and left to undergo *very gradual
staining*, a process which may occupy eight hours, or even
longer. When a section removed to the microscope stage
shows sufficient staining, the reagent may be poured off,
superfluous carmine removed by gentle washing, and all the
sections immersed in a dilute solution of glacial acetic acid
(0·5 per cent.) for fifteen to twenty minutes. The acid
deprives the specimen of all diffuse staining, and fixes the
dye more especially in the germinal centres, whilst at the
same time the tint is brightened by its agency. Great care
should be taken that neither section nor staining solution is
very alkaline in reaction, otherwise the staining will prove too
deep and diffuse, and the structure itself injuriously affected.
A minimum amount of ammonia, however, must be present
in the dye *just recognizable by its odour*, and if this be not the
case, a drop or two of the liquor ammoniæ should be added
to Beale's strong dye prior to making the dilute solution.
These sections may now be washed and mounted in glycerine
or Farrant's solution, or they may be dehydrated in the usual
way by spirit, cleared up in oil of cloves, and mounted in
Canada balsam or dammar.

With a few of these carmine-stained sections the student
should try the following modified process of mounting,
whereby valuable information may be obtained. Several
years ago I described a method of displaying a great wealth
of structure on the cortex cerebri by altering the refractive
indices of the structural elements, and so producing very
remarkable differentiation of structure.[1] I there stated what
follows :—"On placing an unstained section of cerebrum or
cerebellum saturated with spirit in the field of the microscope, little or no structure is apparent, but if a drop of
essential oil be now allowed to run over it, there will be
observed at a certain stage of clearing up, and whilst the spirit
is evaporating, a sudden starting out in bold relief of the
cells, nerve-fibres, blood-vessels, etc., which again disappear or
partially fade on perfect clearing of the section. Now, this

[1] *Quarterly Journal of Micros. Soc.* vol. xvi.

appearance may be fixed by suddenly dropping over its surface a little balsam and permanently mounting. Upon this fact depends the process now to be described. Sections treated with Beale's carmine solution (1 to 7) and washed with the acid, are placed, saturated with spirit, upon a slide. When the spirit has nearly all evaporated, a drop of oil of anise is allowed to flow *over* the section (not to float it up), and the clearing is watched on the stage of the microscope; then, just when the appearance referred to above is presented to view, a drop of balsam is allowed to run over the section, and a covering glass permanently fixed on. In lieu of oil of anise, I frequently employed glycerine with the same results, mounting the specimen in glycerine jelly. * * * I will state my belief that this method will yet prove of most essential service in the estimation of the relative proportion of cell-processes in any individual section, and the most accurate tracing of any existing connections, for not by the deepest aniline staining have I yet succeeded in demonstrating the existence of so thick and numerous a series of processes diverging from the pyramidal layers of the cerebral cortex as by the method described above." It is satisfactory to find these opinions confirmed by Prof. Stirling in his recent text-book of "Practical Histology." Dr. Stirling says :—" The processes of the cells (of the cerebrum) are best seen in preparations which are only *partially* cleared up under the influence of clove oil. This is a most important method of investigation. I have often seen in this way delicate fibrils, not unlike elastic fibres, and which are not distinct when the section is completely cleared up."[1]

3. **Notes on the Reaction of the Dye.**—Carmine is especially suited as a dye for the large nerve-cells, their contents, nuclei, and nucleoli. It exhibits well the connective cells and the vascular apparatus. For the large ganglionic cells and their immediate environment, nothing can perhaps surpass a successful carmine preparation, whilst it does not

[1] "Text-book of Practical Histology," p. 101. Smith, Elder, & Co.

appear to produce any further shrinking of protoplasm or connective such as is produced by hæmatoxylin. It is less adapted for displaying the cell-processes, and wholly fails to exhibit the details of structure in smaller nerve-cells, such as those of the upper cortical layers. It has been regarded, even by so eminent an authority as Charcot, as the only reliable dye for exhibiting the morbid state regarded as sclerosis of the different columns of the cord. This, however, is not the case, since aniline blue-black portrays the lesion even more distinctly. The more important objections attached to this dye are the variable results given ; the little definition often afforded of the tissue elements from the tendency to diffuse staining ; the unpleasant glare of the carmine tint for continuous work, and its unsuitability for examination by artificial light.

PICRO-CARMINE STAINING.

1. **The Dye.**—This valuable reagent, introduced by Ranvier, is of very special use in the preparation of nervous tissues. It is readily made as follows :—

RANVIER'S FORMULA.

Best carmine	... ...	1 gramme.
Water	... ...	10 c.c.
Liquor ammoniæ	... ...	3 c.c.

Rub the carmine up with water, add the solution of ammonia to the fluid in a test tube, and aid the solution of the carmine by gentle heat carefully applied. When dissolved and the solution perfectly cold, pour it into 200 c.c. of a saturated aqueous solution of picric acid. Place the solution in an open vessel and with gentle warmth evaporate to one-third of its bulk. Filter and keep in a stoppered bottle labelled " Ranvier's picro-carminate dye."

Picro-carmine may be purchased in a solid form as a granular and imperfectly crystalline substance—a 1 per

cent., solution is then used. From personal experience I
cannot commend its use, and I would strongly recommend
the preparation after the formula above given.

2. **The Staining Process.**—Immerse the sections in a
small quantity of the picro-carminate solution for a period of
from twenty to thirty minutes. Pour off the dye, draining away
all the superfluous fluid, and then, *without washing the sections,*
float them up by glycerine in which they must be mounted.
It will often be found advisable to stain each section upon
the slide, dropping over sufficient of the reagent to cover it—
to clear it up with glycerine and mount it permanently upon
the same slide. Picric acid, being soluble in water, is readily
removed upon washing these sections; hence, to preserve the
colour due to both pigments, we must carefully avoid washing
with water.[1]

Upon the other hand, washing freely, dehydrating with
spirit, and mounting in the usual way in balsam, will yield
us excellent carmine preparations, and so far as my
experience goes, better specimens than those subjected to
the action of the simple ammoniacal or borax solutions.

3. **Notes on the Reaction of the Dye.**—Picro-
carmine has the special merit of showing less tendency to
diffuse staining than simple carmine dyes. The presence
of the two pigments appear notably to restrict their action
to the germinal centres, for which they each have affinity.
Elastic tissue and muscle will be stained yellow, whilst nuclei
and connective take the carmine tint. Minute nervous
plexuses are poorly shown, as they occur in the structure of
the cortex; hence, the nerve-cell processes and their ramifications can rarely be followed to any distance. The structure
of the investing membranes and the vascular apparatus is
better defined by this reagent than any other we are acquainted
with. In fact, the merits and demerits of carmine staining,
pertain to the picro-carminate, although, for powers of

[1] This difficulty is also overcome by washing in a saturated aqueous solution of picric acid, and dehydrating in an alcoholic solution of the same
acid.

differentiation, it is superior to simple ammoniacal carmine, and its *rôle* as a staining reagent more extensive. For combination with other dyes, such as aniline, logwood, and osmic acid, it is peculiarly useful. As a general dye for nerve-structures, it ranks next to aniline and aniline picro-carminate, and is far superior to logwood.

ANILINE STAINING.

1. **The Blue-black Dye.**—This, which is by far the most valuable of the aniline series of dyes for the brain and spinal cord, is thus made :—

Aniline blue-black	1 gramme.
Distilled water	400 c.c.

Dissolve, filter, and keep in a stoppered bottle, labelled " aniline black dye, aqueous solution, 0·25 per cent." It may, however, be convenient to keep a stock solution of 1 per cent. strength, as rapid staining is occasionally required, and to dilute, as required, to 0·25 or 0·5 per cent., always filtering these dilutions prior to use. The solution, as above given, may be employed for staining fresh and hardened sections, the aqueous solution being the best dye we possess for the former. The alcoholic solution, as first recommended by Mr. Sankey, was the following :—

Aniline blue-black ...	5 centigrammes.
Water	2 cubic centimeters.

Dissolve and pour into it 99 c.c. of methylated spirit. Filter the solution, and label "alcoholic solution of aniline black, 0·05 per cent."

Professor Stirling advises the use of a solution double the strength of the above, thus :—

Aniline blue-black ...	1 decigramme.
Water	4 cubic centimeters.

Dissolve, and add 100 c.c. of rectified spirit, and filter. Label, " aniline blue black, alcoholic solution, 0·1 per cent."

I much prefer the aqueous solution for most purposes.

2. **The Staining Process.**—Sections of hardened cortex, pons, medulla, and spinal cord, may alike be left in the aqueous solution (0·25 per cent.) for one hour. In this time they will, generally, be found stained to a sufficient extent, when they must be removed to a vessel containing water, and well washed from superfluous dye. Next, dehydrate by rectified spirit or absolute alcohol; clear up by oil of cloves, and mount in balsam or dammar.

Sections of Cerebellar Cortex should be deeply stained by aniline, gently washed in water, and immersed for twenty to thirty minutes in a solution of chloral hydrate (2 per cent.). Next transfer them to the following solution :—

Solution of chloral (2 per cent.).
Oil of cloves ...   .   equal parts.
Alcohol   ...   .   q. s. to dissolve perfectly, and form a clear solution. Add the alcohol by degrees, stirring the solution with a glass rod, and avoiding excess of spirit. During use, carefully cover the watchglass containing this solution, so as to avoid evaporation. The chloral removes diffuse staining, whilst the clove oil clears up the section, and enables us, by examining one occasionally under a low magnifying power, to decide when they have reached a satisfactory stage. When this has been attained, remove your section to a slide; rapidly wash with a little pure alcohol; thoroughly clear with clove oil, and mount with balsam. The alcoholic solution of aniline also stains sections of cerebrum, pons, and medulla, and must be cleared and mounted in the usual way. Glycerine being a powerful solvent of aniline, must not be made the medium for mounting these sections.

3. **Notes on the Reaction of the Dye.**—Its action is energetic, and its results are certain and constant. It is readily controlled, so that any depth of staining may be obtained with ease. The tint given varies from a bluish-grey to

a deep blue-black, and is a pleasant one for the eye, causing little or no fatigue. It enables us to obtain the clearest and sharpest definition of elements in a tissue, without modifying their structure by shrinking or other change. No other staining reagent displays the nerve-cell processes to such a remarkable extent as does this dye, the very finest ramification being followed out with ease. It has a special affinity for protoplasm, staining the nuclei of the nerve-cells most deeply, then the nerve-cells themselves and their protoplasmic extensions, and, to a less intense degree, the nuclei of the neuroglia and perivascular walls. The finely-formed nerve and connective mesh-work, forming the basis substance of the cortex, is stained of a pale grey. This dye fails to produce any action upon the medullated sheath of nerve-fibres, affecting only the axis-cylinder. Transverse sections, therefore, of nerve-fibres, such as are seen in cross sections of the medulla, cord, and nerve-trunks, exhibit the axis-cylinder stained of a bluish-black tint, the unstained white matter of Schwann surrounding it ; whilst the intertubular connective, lightly stained, differentiates the medullated tubes from each other. The value of this dye for connective tissue is very limited, its *rôle* being chiefly confined to the nerve-elements.

OSMIC ACID STAINING.

This method we owe to Professor Exner. It is peculiar in the fact that the hardening and staining proceed simultaneously, the same reagent being employed for both purposes. The preparation of the hardened brain by immersion in osmic acid has already been alluded to, and we have now simply to detail his further treatment of the sections so obtained by the microtome. The sections, owing to the deep staining they undergo, must be *extremely thin* and should be placed in glycerine immediately, since prolonged immersion in alcohol injures them seriously. Transfer a section to a glass slide and add a few drops of strong liquor ammoniæ. Absorb all superfluous moisture around the section

with blotting paper, and after waiting a short time for the
ammonia to effect the section thoroughly, place the cover
glass on without injury to the preparation which has been
softened by the reagent, and examine immediately on the
microscope stage. It is at this early period that the section
shows to greatest advantage. For permanent preservation
Exner surrounds the cover glass with the soluble silicate
called water glass. Formerly an ammoniacal carmine solution was used, but Exner found that the results obtained
could be attributed only to the reaction of ammonia, and so
he has dispensed with any carmine staining as the osmium
suffices for all purposes.

Notes on the Value of this Method.—The medullated
fibres can be traced upwards into the highest realms of the
cortex, exhibiting a wealth of structure which no other
method displays. The medullated sheath is deeply stained
and can be traced distinctly in the case of the smallest nerve-
fibres. The nerve-cells, although stained, do not form such
prominent objects as by other methods of staining; this
process being especially adapted for tracing the ultimate
course of medullated fibres through the different layers of the
cortex and determining their final destination. Exner states
that the staining of the minute fibres occasionally disappears
in time, and the student should also be aware that the brain-
structure swells up so much under the agency of ammonia
that the section is increased by one-third its full diameter.[1]

ON THE USE OF COMPOUND PIGMENTS.

Under this head I would include some of the most valuable methods we possess for delineating the minute structures
of the nervous system. Picro-carmine, as consisting of two
pigments, would naturally fall under this category, but since
it is employed in the same way as the simple dyes, I have

[1] For results obtained by Prof. Exner with this method, see Original
Memoir (op. cit.), or abstract of the same in *Brain*, part xv.

found it more convenient to include it amongst them. The methods of staining to which I now refer are those in which the tissue is first submitted to one dye, and the result subsequently modified by the addition of a second pigment. In these cases, either different histological elements assume the tint of each pigment, or the former staining is greatly improved in character by the second pigment combining and modifying the tint acquired. In the former case only does a genuine double staining occur; in the latter we obtain a single tint, but one of valuable quality.

1. **Aniline Picro-carminate Staining.** — Three pigments are here used, viz., the aqueous solution of aniline blue-black (0·25 per cent.), and Ranvier's compound dye, picro-carmine. Sections, which should be very thin, may be stained on a glass slide, by dropping over them from a pipette sufficient picro-carmine to cover the section completely. When deeply stained by this reagent the latter is drained off and the section covered in a like manner by a quantity of the aniline blue-black solution. It will be found that the action of the aniline proceeds much more rapidly upon a section stained by picro-carmine—ten minutes usually sufficing to produce the desired effect. This result is acquired when upon draining off the dye the section is found to have changed from the bright carmine to a deep violet tint. Wash the sections well in water, dehydrate by spirit, clear by clove oil, and mount in balsam. In this way the most beautiful effects are obtained, which are specially valuable and instructive. The tint is much less fatiguing to the eye than the bright glare of carmine, whilst in the differentiation of nervous tissue in the hardened brain this method of staining appears to me unrivalled. The deeper the tint desired the longer immersion in aniline must be practised, but for the more satisfactory and pleasing effects I have usually found it suffice to stain for half an hour in picro-carmine and fifteen minutes in aniline.

Notes on the Reaction of the Dye.—This compound

K

staining enhances the value of an ordinary picro-carminate preparation in the fact that a better differentiation of structure is obtained, and examination by artificial light rendered as agreeable and valuable as by day-light. With all the advantages of a picro-carmine staining we have conjoined the special action exhibited by aniline for the cell-processes and finer nervous meshwork, whilst at the same time the general tint is deeper than by picro-carmine, and the contrast betwixt the deeper and lighter stained parts more marked. The neuroglia is stained faintly; the connective nuclei are prominent objects; the nerve-cells are well shown, the nucleus taking a much deeper tint. The axis-cylinder, in transverse sections of medullated fasciculi, is stained of a deep purple, the intertubular connective and its nuclei being of much lighter hue. The vascular channels and their nuclear elements, which are not satisfactorily shown in simple aniline staining, are exhibited to great advantage. I find this method of staining very suitable for displaying the coarse structure of the brain, and sections through a whole hemisphere treated by this method are most instructive for naked-eye examination or by low-power objectives.

2. **Picro-Aniline Staining.** — A useful staining of medulla and spinal cord may be obtained by employing Judson's " Cambridge blue " as the first pigment, washing off superfluous dye, and then by momentary immersion in a saturated solution of picric acid the section acquires a brilliant green, which exhibits structural details remarkably well. Since both pigments are soluble in alcohol, the sections should be very rapidly dehydrated, cleared by clove oil, and mounted in the usual manner.

3. **Osmic Acid and Picro-carmine.**—The section is immersed in a solution of osmic acid (1 per cent.) carefully covered over so as to confine this very volatile reagent as far as possible to the tissues to be acted upon. When the latter has acquired a dark-brown tint the section should be removed, well washed, and lightly stained with picro-carmine.

Such sections should be mounted in glycerine or glycerine jelly. Sections through the hemispheres of the brain in small animals, or the pons and medulla may be conveniently treated by the osmic acid contained in a watch-glass covered by a small glass plate. The above is a most valuable method of staining.

4. **Hæmatoxylin with Aniline.**—This, like the last, gives a genuine double staining. Sections are first stained by logwood according to the directions already given, and immersed *for a few seconds only* in the aqueous solution of aniline blue-black (0·5 per cent). A solution of chloral hydrate (2 per cent.) will remove any excess of aniline staining. Wash the sections well, *rapidly* dehydrate by alcohol, and mount *sec. art.* This process is chiefly adapted for sections of cerebellum, and its special value will be referred to later on. The above are all the methods of staining I have found to be of greatest value, but I append here two methods of double staining for the cortex of the cerebellum, advocated by Professor Stirling,[1] of the merits of which I have not had personal experience.

5. **Eosin and Logwood.**—Stain a section for a few seconds in a very dilute watery solution of eosin (1 part to 1,500 of water) until it has a faint red colour. Great care must be taken not to over-stain. Wash with water, stain the section with logwood, and mount in dammar.

6. **Picro-carmine and Iodine Green.**—Stain a section in picro-carmine. Wash it in water acidulated with dilute acetic acid, and, after washing in pure water, stain again by means of iodine green, and mount in dammar.

§ STAINING OF SPECIAL REGIONS OF THE BRAIN AND SPINAL CORD.

The Cortex Cerebri.—The cortex at the vertex in the frontal regions should by preference be stained by the

[1] "Practical Histology." p. 100.

aqueous solution of aniline blue-black which has so remarkable an affinity for the nervous protoplasmic masses and their
extensions, and which in these regions are so numerous.
Next in value to aniline are the picro-carminate and logwood
dyes. With care very beautiful sections may be obtained
from this part of the brain stained with picro-aniline, but a
good aniline blue-black surpasses all others in intrinsic merits
and for purely anatomical details. When it is desired to
examine critically the contents of the nerve-cells with regard
to morbid change the picro-carminate gives equally good
results with the aniline. No reagent, on the other hand,
exhibits so remarkably the proliferation of Deiter's corpuscles
in the cortex as does aniline.

In more posterior planes, and especially at the occipital
pole of the hemisphere, where layers of the granule-cell
formation predominate, the hematoxylin dye brings out the
distinctive features of these realms better than any other.
The nuclei of the crowded granule series are deeply stained
by this reagent, although the large nerve-cells in the intermediate layers are but faintly seen. It is on this account
advisable to attempt double staining by logwood and
aniline when both granule cells and the larger nerve-cells
are well shown. The same remarks apply to the cortex in
the olfactory region, where, in like manner, small granule
elements predominate.

The peculiar crowded cell-formation of the cornu ammonis
stained by aniline blue-black affords beautiful preparations.

The Cortex Cerebelli.—The granule layer of the
cerebellar cortex is readily stained by any of the pigments
more generally used, but not so the layers of Purkinje or their
cell-processes, which branch out so abundantly in the pure
grey layer.

Picro-carmine and logwood alike yield fine, yet imperfect,
preparations, since the mere outline of the Purkinje cell or
its nucleus is often all that is visible in this layer. Occasionally, however, a few branches may be seen. To exhibit

the great profusion of branches arising from these cells stain with aniline black, or preferably employ the double staining by logwood and aniline black. In these cases the granule layer will be of a bright purplish blue, whilst the cells of Purkinje will be distinctly mapped out, the nucleus of a bluish black tint, and their branches displaying a perfect forest of twigs.

The Central Medulla.—The student should make sections, both in a vertical and horizontal direction, through the brain of the smaller mammalia, *e.g.*, the cat, dog, or rabbit, and stain these preparations with the double object of utilizing them for coarse examination by the naked eye or two-inch objective, to learn the relationships of the various central structures at different planes, as also for the more arduous duty of tracing out the course taken by the different systems of medullated fasciculi in the interior of the brain. Such preliminary work he will find invaluable when he attempts to unravel the intricate mechanism of the human brain.

For the coarse examination of the larger sections through the whole brain or one hemisphere, the aniline picro-carminate has always appeared to me the better method of staining, but for the more minute microscopic tracing of central fasciculi the double staining by osmium and picro-carmine is the only reliable method.

It must be remembered that whenever we are dealing with sections carried transversely to the axis of the medullated fasciculi these latter will be well delineated by aniline blue-black, picro-carmine, or logwood; but where the plane of section lies parallel with the long axis of these fasciculi, the more reliable staining is that of the medullated sheath by means of osmic acid. Hence it arises that in following out the lengthened course of fibres through the brain by sections along their planes, osmium is the most valuable adjunct, whilst the structure of the medulla, studied in transverse section, is best treated by other reagents. Although when dealing with the more extensive areas, as in sections through the hemispheres, it is advisable to obtain extremely thin preparations, yet the vast majority of valuable sections for the purpose of tracing the central medulla are most unsuitable for minute examination of the cortex, the details of which will be very poorly delineated. The choicest sections through the cortex, such, in fact, as are essential to a favourable investigation of its structure, must be of extreme tenuity—a degree indeed not attainable in large sections through the hemispheric mass.

Let the series of preparations illustrative of the minute structure of the cortex in different realms, therefore, be wholly distinct from those illustrative of the central medulla and basal ganglia, since the methods of preparation which are most suitable for these diverse parts differ so essentially. On the other hand, Exner's process is available for both regions.

The Pons, Medulla, and Spinal Cord.—The same method of staining is applicable to all these parts. Logwood has been used extensively for these regions, carmine still more so ; whilst the aniline reagent, which I regard as of far the greater value, has been comparatively neglected, at least so I judge from the fact that whenever sections of cord or medulla are exhibited, the carmine preparations infinitely outnumber the aniline. A tolerably thick section of medulla or cord appears wonderfully thin when stained by carmine and cleared up by oil of cloves ; it shows to very fair advantage even to the higher powers of the microscope, and any inequality of surface or scoring from the frayed edge of the section-knife is obscured or entirely concealed from view. This is not the case with the aniline preparations. The depth of tint here not only warns us but too surely of any undue thickness of our section, but shows up any faults in the section due to scoring or tearing of structure to a remarkable and most unpleasant degree. Hence, I incline to the belief that faults really attributable to the section-

cutter and his blade' have been credited to the dye, which
has, therefore, fallen into comparative disrepute. Of this,
however, we may be certain, that sections of medulla and
cord properly stained by aniline blue-black are the most
beautiful preparations of these regions we can by our present
methods possess; and that a really good aniline preparation
of these parts must of necessity be a *very thin and perfect
section*,—a statement which can scarcely be made of carmine
preparations. The student should therefore possess himself
of the thinnest possible sections of the different regions of
the medulla and spinal cord, prepared by the aqueous
solution of aniline blue-black, and with due care to the condition of his section-knife, he may thus prepare a most
invaluable series of sections; yet from the necessary fine-
ness of the sections, he will meet with greater difficulty in
obtaining perfect preparations than by the carmine staining.
Whilst thus giving the foremost place to aniline blue-black
for staining these regions, I would not wish to undervalue
the merits of picro-carmine. Next to aniline, it is pre-
eminently useful here. Do not, however, be deceived by its
peculiar power of concealing the demerits of your section,
and so be led to regard a section as thin because it appears
so when mounted. In this way, minute details may be
wholly overlooked which really thin sections would have
exhibited. With the high powers the actual thickness of
the section may be decided readily.

APPARATUS REQUIRED FOR STAINING AND MOUNTING.

Bottles of staining reagents	Glass rods
Porcelain capsules	Pipettes
Watch-glasses	Camel-hair brush
Slides	Funnel and filtering paper
Cover-glasses	Mounted needles
Tray for fresh slides	Turn-table
Bottles of mounting media	Bottles of cement

CHAPTER X.

AMONG the facilities offered for *immediate examination* of brain and spinal cord in the recent state, two processes present themselves for examination ; these are—

1. Methods of section-cutting and preparation of frozen tissues.
2. Methods of dissociation and teasing out of fresh tissues.

We should perfectly familiarize ourselves with all these methods. There are certain difficulties peculiar to the preparation of fresh nervous tissues, and not met with in the manipulation of hardened sections, with which it is well to be acquainted. We shall recognize in the details of preparation now to be given, the means adopted for overcoming these obstacles.

Sections of frozen brain must be extremely thin for satisfactory preparations, and these most delicate films absorb water so rapidly that the constituents undergo rapid dissociation when floating in water. Again, after removal to a slide, the upper surface is found highly repellant of watery solutions, so that the application of osmic acid equably over its surface is a matter of some difficulty when dealing with so fine a film. Lastly, fine sections of spinal cord exhibit such rapid endosmose of fluid, that unless each section be removed immediately it is floated off the blade into glycerine, the nervous elements swell up and burst through the encircling girdle of membranes, becoming completely dissociated. The first difficulty as regards the cortex is met by the osmic acid treatment, the second is overcome by a little practical manipulation, and the difficulty connected with section of the cord is wholly avoided by glycerine diluted with one-half or two-thirds its bulk of water.

It must be remembered that all *alcoholic* and *highly astringent reagents* must be excluded from our list for staining purposes, and therefore hæmatoxylin and several other dyes employed for hardened brain are wholly inadmissible for fresh nervous tissues. We are, therefore, chiefly restricted to aniline blue-black, carmine, picro-carmine, and osmic acid. The osmium and aniline are especially valuable agents.

§ PREPARATION OF SECTIONS FROM FROZEN TISSUES.

The Freezing Method.—Use the ether freezing microtome and secure it to a firm support, as a table or ledge in front of an open window, where a cool draught plays over the section-plate. Place a large deep glass vessel half-filled with clear water on the support in front of the microtome ; have the ether-spray apparatus ready on the left hand, and the section-blade on the right. Throw across the knees a soft towel, with which the *under surface* of the section-blade has to be gently wiped occasionally to remove superfluous moisture. Next take up the section-blade, and holding it horizontally, pour upon its *upper surface* a small quantity of ether, allowing it to sweep from end to end, and agitate the blade in water until the blade, on removal, remains perfectly wet from end to end. Repeat the ether wash if the surface still remains somewhat greasy, as in this state it is unsuited for section-cutting. Lower the freezing chamber of the microtome until its surface, or freezing plate, is four or six millimeters below the upper surface of the section-plate. Place upon the upper surface of the freezing chamber the tissue to be frozen, which should be but a little over a quarter of an inch thick. Introduce the nozzle of the ether-spray instrument through the opening in the left-hand of the freezing chamber, so that the spray be directed upwards against the freezing plate immediately beneath the tissue. A steady current will, in a few seconds, cause blanching of the tissue next to the plate, to which it now firmly adheres, and freezing proceeds rapidly upwards, as evidenced by its colour and firmness. It is not necessary to freeze above the level of the section-plate. Remove the unfrozen tissue down to a level with the section-plate by a sweep of the blade—plunge the latter into water, lightly wipe its *under surface* upon the towel, and raising the section by the micrometer-screw, cut the first section with a sweep away from you, and from your left to your right hand. The section must be *floated off* in the glass vessel, and *not plunged beneath* the surface of the water.

One should aim at acquiring the greatest expedition at this work, rapidly floating off each section, and wiping the blade preparatory to the next. It is all-important that fresh sections be removed from the water in which they float as quickly as possible, whilst, at the same time, a number of sections should be cut before the frozen mass begins to thaw and loosen from the plate. Practised manipulators can with case cut a dozen fine sections ere the mass loosens from the plate, but the tyro, when simply practising the cutting of sections, is slow in his manipulation, and finds the mass loosens before he has secured three or four specimens. Two or three puffs from the spray, however, will, at this stage, fix it to the plate again, and thus he may with ease cut through the whole depth of tissue with very little expenditure of ether. When he has obtained sufficient manipulative skill at section-cutting, he will find it advisable to pass the first half-dozen sections through the second stage of the process before he cuts another series, so as to avoid the injury resulting from prolonged floating in water ; this last statement holds good for all nervous tissues alike. There is, perhaps, no method of obtaining fine sections of tissue so expeditiously and with

such extreme simplicity of detail as that just described, and
certainly none involving so little expenditure of time or labour ;
yet, as the beginner will always meet with some obstacles in
his first attempts, however simple the process, it will be as
well to indicate those which most frequently present themselves. His failures will invariably be due to one or other of
the following causes :—

1. *The temperature of the room above* 60° *F.* This also
involves unnecessary waste of ether. Freezing should only
be practised by ether spray when the temperature of the
atmosphere falls below 60° F.

2. *The brain mass frozen too high.* Brain-tissue, when
frozen hard, cannot be properly cut ; in fact, our sections are
invariably obtained from the tissue which is slowly thawing,
and which is always of most delicate consistence for cutting.
On this account it is not necessary to freeze up to the level of
the section-plate.

3. *Too much water on the blade.* If this superfluous fluid
runs off on to the section and freezing-plate, it sets into a
solid icy mass, which injures the blade and greatly retards
section-cutting.

4. *Water freezes over the section-plate.* The uneven surface thus produced prevents section-cutting. The difficulty
is removed by never neglecting to draw the under surface of
the blade over the towel before each section is taken.

5. *Brain freezes too hard irrespective of the above conditions.*
This only results when the tissue is extremely œdematous
from morbid states, and when the temperature of the surrounding atmosphere is so low that freezing is unusually
rapid. The difficulty is met by moistening the upper surface
of the blade with glycerine instead of water, and carefully
guarding it from extending to the freezing plate.

6. *Sections tear or cling to the blade.* This results from
the greasy surface of the blade retaining no water to float up
the section.

7. *Tissue cannot be frozen to plate or loosens too soon.* The

first occurs when glycerine or other medium which cannot be
frozen has been accidentally smeared over the plate—the
latter usually from want of rapid manipulation on the part of
the operator, or too warm an atmosphere for satisfactory
freezing.

8. Lastly, *the well or freezing chamber may be too small*,
in which case freezing is not only retarded, but subsequent
thawing is hastened. For my own part I would strongly
recommend this portion of the instrument to be full 2 inches
in diameter by 1½ inches deep.

We now proceed to the subsequent treatment of our
sections.

Staining Fresh Sections of Cortex.—In this stage our
sections pass through two processes—the osmic acid treatment and aniline staining. Take a perfectly clean glass
slide, pass it beneath a section floating in water, and, fixing it
to the slide by a needle, remove it carefully, draining off all
superfluous fluid from the slide. Remove each section in a
similar way, and next, by means of a pipette, let two or three
drops of osmic acid solution (0·25 per cent.) fall on the slide
and float up the section; with a pen-knife carefully spread
this fluid over the upper surface of the section, which, on
account of the repulsion afforded, gives a little trouble with
extremely thin films. This procedure is, of course, not
necessary where the section has been sunk in water, but the
more delicate sections float, and can be plunged beneath the
surface of the water only at the risk of ruining them. A
little practice soon enables us to carry the osmic acid over the
most delicate films without tearing them. Sections thus
treated should not remain over one minute in the osmic acid
of the strength named, but should at once be immersed in a
vessel of pure water and left here for five or ten minutes. By
this time all superfluous acid has been removed, and the
osmium has "fixed" the nervous structures, so that they are
manipulated without danger and do not deteriorate in water.
The sections may now be stained collectively, or, as I prefer

it, each section should be placed on a slide, and sufficient aniline blue-black dye (0·25 per cent.) poured on to completely cover the film. Place the slides under cover. It is well to have a small tray with a shelf attached around its sides on which the slides may rest, the tray being fitted with a cover. The staining with the aqueous solution of the dye occupies about one hour; alcoholic solutions of the dye are of course inadmissible. When staining is complete a pipette will remove superfluous dye, or it may be drained off carefully, and the sections again plunged into water and gently moved about until well washed from unfixed dye. There are certain circumstances which the student should be aware of in adopting this process of staining, and these I will here briefly allude to.

If the osmic acid solution be too strong or left on the section too long, subsequent staining by aniline will prove very unsatisfactory, and, moreover, the osmium staining becomes apparent, and is dissolved out subsequently by the benzole solution of balsam tinting the latter strongly and spoiling the preparation. Since the osmic acid is used simply to give the sections the necessary consistence for manipulation, and not for the purpose of staining, it should be employed in the weakest form consistent with this action, and the sections exposed to its agency for the very shortest period necessary, and very thoroughly washed prior to staining. If the osmic acid be unequally spread over the section, those parts not immersed will stain with less vigour by aniline.

Again, when all the above precautions have been taken, it is sometimes found that sections exhibit a rough, irregularly-stained surface, really due to the edge of the section-blade being somewhat frayed, and so tearing the delicate structures in cutting through them.

Mounting for Permanent Preparations.—Let each section, when well washed, be received upon the centre of a perfectly clean glass slide; all fluid drained off, and the specimens again placed upon the shelf of the covered tray, to

dry spontaneously. The process may be hastened, if necessary, by enclosing them within a bell-glass along with a
capsule containing strong sulphuric acid. When perfectly
dehydrated, let a drop or two of a benzole solution of balsam
fall on the section, and adjust the cover-glass.

Staining Fresh Sections of Central Medulla.—
As before stated, the same method of staining is not alike
suitable for both cortex and medulla. The aniline process,
just described, is in every way adapted for the cortex of the
cerebrum and cerebellum; but when our object is to follow
the course of the central medullated fasciculi, and examine
the structure of the basal ganglia, the use of osmic acid has
to be invoked. Before practising on the more extensive areas
of the human cerebrum, it would be well for the student to
slice the brain of one of the smaller animals—(the cat, dog,
rabbit, are very suitable)—upon the freezing microtome,
carrying sections through the hemispheres in vertical and
horizontal planes. Such sections should be extremely fine,
and on removal on a slide to the stage of the microscope,
should exhibit, in an exquisite manner, the details of
medullated structures. The grey matter will, of course, be
translucent, and apparently structureless; but the medullated
fasciculi will appear in striking relief of a pure white, the
minutest fibre readily followed by a comparatively low-power
objective. Our object, however, is to render these structures,
not only clearly-defined, but permanent preparations, and this
may be effected by the following procedure. The finest
sections are removed *as rapidly as possible* from the water in
which they float upon a slide, and floated up immediately by
osmic acid (1 per cent. solution), drawn also carefully over
the upper surface of the section. The slide is placed aside,
closely covered over, so as to confine the vapour of this very
volatile reagent as much as possible to the section. In a
period, varying from five to ten minutes, the sections are
floated off into water, and freely washed. They acquire
rapidly the deep brown tint of osmium, and can be manipu-

lated without fear of tearing. The depth of tint required can be best learnt by experience.

After well washing, the sections may either be mounted as they are, or immersed for a few moments in picro-carmine; the latter is, perhaps, the best plan to adopt, as the grey matter is thus shown stained, in striking contrast to the medulla.

The mounting of these preparations is more tedious, as they require to be preserved in glycerine or aqueous solutions, and careful sealing-up is therefore requisite. Take up the sections on the centre of a slide, after washing off the picro-carmine dye, and allow a drop or two of strong glycerine to float up the sections and remain exposed for five or ten minutes. The section becomes clearer and freed from much water by this means, and the glycerine is drained off. Just sufficient strong glycerine is again dropped on to the section to cover it completely, and the cover is placed on. All traces of superfluous glycerine must be carefully removed from the margins of the cover-glass after it has been gently pressed down, and on no account must any glycerine remain on the slide just outside the cover-glass. A thin coating of the lead cement is laid on around the square or circle of thin glass, just overlapping by a couple of millimeters the edges of the latter. This must be allowed to dry completely, when a second coating is run on, and also allowed to dry. Over this, a layer of gold size may be painted, and a finish afforded by the white zinc paint. This method of sealing-up, although tedious, is most reliable and durable; it well repays the trouble taken. The same method may be employed where preservative fluids, such as Farrant's, Goadby's, or sodium chloride with camphor solution are used. We should be cautious in using these fluid preservatives, to lay the gold size *above and not beneath* the red lead cement. A weak solution of chloride of sodium, made with camphor-water, also answers well for preserving the preparations stained by osmic acid alone.

§ PREPARATION OF NERVOUS TISSUES BY DISSOCIATION.

Prior to the systematic section-cutting of frozen brain, various methods of fresh preparation had been tried. Thus, a thin section of cortex was often pressed out between the cover-glass and slide, and its vascular system fairly well displayed; but the nerve-cells and their delicate ramifications could not, of course, be detected by such a coarse method of procedure. It was found necessary to prepare the section by certain reagents whereby dissociation of elements might be secured, whilst some form of staining reagent was likewise essential. Amongst the earlier attempts of this kind, was that made by Gerlach,[1] but the process is applicable only to the large ganglionic cells of the spinal cord. I have myself employed a process less tedious than that of Gerlach, having the advantage of being rapid and immediate in application, with results which, to me, appear equivalent to what is obtained by the more lengthened process. I shall describe both.

Gerlach's Method of Dissociation.—The finest possible sections are made longitudinally, through the perfectly fresh spinal cord of the calf or ox, and are immediately placed in a very weak solution of bichromate of ammonia (1 part to 5,000 or 10,000 parts of water). Let them remain here for two or three days, at a cool temperature. Next, immerse them for twenty-four hours in a dilute solution of ammoniacal carmine. Wash the sections well with distilled water, and tease out the nerve-cells with a needle, under a lens. Preserve in glycerine, or evaporate to dryness on a slide; moisten with oil of cloves, and mount in Canada balsam.

Dissociation may thus be practised in the case of ganglionic cells, wherever they are found, the macerating fluid being either dilute bichromate of potash or ammonia, Müller's fluid,

[1] Stricker's "Human and Comparative Histology." Syd. Soc. Vol. ii. p 345.

or iodized serum, the process of maceration occupying from one to three days.

Rapid Dissociation and Teasing.—Make a fine section with a razor through the anterior cornu of the cord in its lumbar enlargement. The section must be made longitudinally, and either by hand or on the freezing microtome. Place the section in the centre of a glass slide ; saturate it by immersion in Müller's fluid for a few minutes ; place a glass cover over the section, and gradually compress it by a mounted needle between the cover-glass and slide. Before a fine film is obtained, the slide is carefully wiped on its under surface, and examined by a low power under the dissecting microscope. Even as opaque objects, the ganglion cells will now attract our attention by their little masses of pigment. With the slide still under a low objective, steadily and gradually compress the part where the cells chiefly congregate, and, by a gentle rocking movement of the cover-glass, we may thus spread them out over the field, or aggregate towards the clearer areas, whilst any alteration they might suffer from this compression, is sure to be very trivial, owing to the elastic resistance of the neuroglia framework. The cover-glass may now be partly pressed and partly tilted off the film, and the mounted needle used to arrange the cells according to our mind, whilst the other textures around may be carefully removed. A few drops of Beale's strong solution of carmine, or the aqueous solution of aniline blue-black (0·25 per cent.), will very rapidly tint the germinal centres of the required depth of hue, and should then be removed by immersion in water, or by allowing a few drops of water gently to flow over the preparation. The cells are dried *in situ*, moistened with clove oil, and mounted in balsam.

Dissociation and Staining of Ganglion-cells of Cortex.—The two methods just detailed are available only for the cells of the spinal cord. For the ganglion-cells of the human cortex, I devised a method which

L.

was described in the *Monthly Microscopical Journal* for 1876, and which affords very satisfactory results. The process may be considered under three stages: 1. Obtaining the film ; 2. Staining; 3. Permanent mounting of the preparation.

1. **First Stage : The Film.**—Strip the convolution of its membranes, and excise a portion convenient for manipulation. Hold it betwixt the thumb and second finger of the left hand, so that the index finger may guide and support the blade. Then, with the razor-blade saturated with methylated spirit, cut section after section vertically to the surface of the cortex, and of the greatest delicacy possible by these means. Very thin sections may thus be obtained with little trouble. Each should be placed on a glass slide, and three or four drops of Müller's fluid allowed to fall upon them from a pipette. Maceration in this fluid for a few seconds, prevents the film adhering, during the next procedure, to the cover-glass or slide. The cover is next placed on, and so arranged as to cover the film by only one-half its diameter. Place the point of a mounted needle upon the centre of the cover-glass, and by steady pressure flatten out the section into an almost transparent film. Practice will soon enable the operator to so arrange the cover-glass that the extreme edge of the compressed section still occupies the space covered by the glass circle. Now remove the superfluous reagent by rapid rinsing in water, and place the slide in a flat porcelain bath or tray, containing methylated spirit. In from thirty to forty seconds, the cover can be removed without tearing the film. For this purpose, remove the slide, steady the cover by the pressure of a finger against one part of its edge, and pass the point of a needle under the edge, near the medullated portion of the section, and *gradually* elevate it. Wash the films from spirit by allowing a stream of water to flow over them from a large camel-hair brush.

2. **Second Stage : Staining.**—The aniline blue-black solution (1 per cent.), is the most suitable staining reagent.

It is dropped upon the film as it lies on the slide, and when sufficiently stained, the superfluous dye is poured off, and the slide lowered gently into water, the film being washed by agitating the fluid around it with a camel-hair brush. Carmine and picro-carmine may be also used for staining; but hitherto the best results have been obtained with aniline black dye.

3. **Third Stage : Permanent Mounting.**—The specimens lying in the centre of their respective slides are now transferred to a bell-glass, so as to exclude dust, and left to dry spontaneously, or the process of desiccation is hastened by including a capsule containing strong sulphuric acid. When perfectly dry, let a drop of chloroform fall on the preparation, and immmediately add the balsam (chloroform or benzole solution), and adjust the cover glass.

EXAMINATION OF NEUROGLIA.

The neuroglia may be well shown in the spinal cord by the methods of dissociation above given. It has been especially studied in the fresh state by Boll, Ranvier, Mierzejewski, and others, and the following methods are worthy of note :—

1. **Boll's Method.**—Thin sections of the brain-substance were plunged into a solution of osmic acid, 0·1 per cent., and after twenty-four hours immersion, were washed in distilled water and kept in a concentrated solution of potassic acetate. For microscopic examination, these darkened sections were dissociated or compressed strongly betwixt cover and slide.

2. **Ranvier's Method.**—Osmic acid (1 part in 300 parts of water) was *injected interstitially* into the fresh spinal cord. The parts infiltrated by the reagent were afterwards removed by the section-knife, torn up, and teased in distilled water, coloured by picro-carmine, and examined in glycerine. This, I may add, is a very excellent and valuable method.

3. **Mierzejewski's Method.**—A fragment of the white substance of the brain, about one cubic centimeter in bulk, was placed for twelve to twenty-four hours in a solution of osmic acid (1 part in 300 parts of water). By this time the preparation was impregnated by the reagent and its exterior dark and callous. The hardened exterior was then sliced off by the section-knife, when the parts immediately beneath it were found pretty equably affected by the osmium. Minute preparations were then obtained from this region, after the manner of Ranvier, viz., by staining with picro-carmine and examining in glycerine. Mierzejewski employed this method satisfactorily in his studies of the neuroglia of the brain in general paralysis.[1]

Dissection of Medullated Strands in the Fresh Brain.—Much valuable information may be gleaned by the student in the examination of the brain of the smaller mammalia by dissection beneath a fixed lens or dissecting microscope. The brain should be placed beneath the light concentrated by a bull's-eye condenser. The scalpel, mounted needle, and camel-hair pencil are all the aids he requires. In fact, the needle and brush do nearly all the work, the broken-down tissue and grey matter being washed off by the brush filled with a solution of chloride of sodium (1 per cent.). It is astonishing to one who has not tried this method how much can thus be done. It is well to begin with the very small brains, such as that of the rat.

Remarks upon Dissociation of Nerve-Cells.—The student must not regard the methods of dissociation just given as being merely a rough and ready means of displaying the coarse structure of nervous tissue. Where the microtome is not at hand, or cannot be employed, these methods are very valuable; but, even when possessed of the permanent sections obtained by the microtome, dissociation should supplement our inquiries. The films thus obtained, as well as those off

[1] "Études sur les Lesions Cérébrales dans la Paralysie Générale." *Archives de Physiologie*, 1873.

the freezing microtome, afford us much information upon histo-chemical examination. The elements may thus be treated by osmic acid, salts of gold, platinum, and silver, and a variety of dyes. For examination of the reactions with chemical reagents, I have found it useful to obtain the faintest tint by aniline, just sufficient, indeed, to bring into view the outline of the cells and their nuclei. For this purpose a few drops of very weak aniline dye, left on for a few minutes, and then washed off, may be fixed in the germinal centres by a drop of extremely dilute hydrochloric acid, which is immediately cleared off by water, and the film left in a satisfactory state for noting the effect of reagents.

PART III

LIST OF REAGENTS AND MOUNTING MEDIA

LIST OF REAGENTS AND MOUNTING MEDIA.

The following list comprises all the more useful reagents which the student will require in the prosecution of the minute examination of the brain. Those marked with an asterisk (*) are the most important, and cannot be dispensed with.

HARDENING REAGENTS.

1.* Chromic acid (stock solution, 1 per cent.).

Chromic acid cryst. ... 10 grammes.
Water ... ... 1,000 c.c.

Dissolve. Dilute to 0·15 per cent. for brain, and 0·25 per cent. for spinal cord.

2.* Bichromate of potassium (stock solution, 4 per cent.).

Bichromate of potassium 40 grammes.
Water (hot) ... ... 1,000 c.c.

Dissolve. Used in dilutions from 0·25 per cent. upwards.

3. Bichromate of ammonium (stock solution, 4 per cent.), made as the potassium solution above.

4.* Müller's fluid.

Bichromate of potassium ... 25 grammes.
Sodium sulphate ... 10 ,,
Water ... ... 1,000 c.c.
Dissolve.

5. Picric acid solution.
Cold saturated solution in water.

6.* Osmic acid.

A solution of 1 per cent. should be kept in a blue glass closely-stoppered bottle, to be diluted when required immediately before use.

7.* Methylated spirit.

8.* Absolute alcohol.

FORMULÆ OF SPECIAL SOLUTIONS.

9. Rutherford's reagent.

Chromic acid ... 1 gramme.
Bichromate of potassium 2 „
Water ... ... 1,200 c.c.
Dissolve.

10. Solutions for Betz's method.

(a) Iodized alcohol.
Alcohol (70 to 80 per cent.), tinted sherry-brown by tinct. iodi. co.
(b) Iodized spirit.
Methylated spirit, tinted sherry-brown by tinct. iodi. co.
(c) Bichromate solutions.
Solutions of potassium bichromate from 0·5 per cent. up to 5 per cent.

11. Solutions for Hamilton's method.

(a) Müller's fluid ... ... 3 parts.
Methylated spirit ... 1 „
Mix.
(b) Solutions (aqueous) of ammonium bichromate, 0·25 per cent. to 2 per cent.

12. Kleinenberg's picric acid solution.
Cold saturated aqueous solution of
picric acid ... ... 300 c.c.
Concentrated sulphuric acid ... 6 c.c.
Mix, filter, and add to the filtrate, water 900 c.c.

STAINING REAGENTS.

The following may be regarded as a fairly complete list of the reagents more generally employed; the formulæ for special solutions are also given below :—

1.* Carmine.
2.* Hæmatoxylin (crystals). Logwood extract.
3.* Picric acid (crystals).
4. Eosin.
5. Gold chloride.
6. Double chloride of gold and potassium.
7. Silver nitrate (0·5 per cent.).
8. Silver lactate (Alferow's formula).
9.* Osmic acid (1 per cent.).

Aniline dyes comprising—

10.* Aniline blue-black.
11.   „   blue.
12. Rosanilin (magenta).
13. Fuchsine.
14. Methyl-aniline.
15. Iodine green.
16. Judson's dyes, especially *" Cambridge blue."
17. Various combinations of the above.

FORMULÆ FOR SPECIAL SOLUTIONS.

1.* Beale's carmine reagent.

Carmine (in small fragments) 10   grains.
Liquor ammoniæ ...   ...   ½ drachm.
Glycerine (Price's)...   2   ounces.
Alcohol   ...   ...   ½   „
Distilled water   2   „

Heat the carmine with the ammonia in a test-tube *carefully*, boil for a few minutes, and expose for one hour. Add the water; filter. Add the spirit and glycerine, and expose until the odour of ammonia is scarcely perceptible.

2. Thiersch's carmine reagent.

(*a*) Carmine   ...   1 part.
Liquor ammoniæ   1 „
Distilled water...   ...   1 „
        Dissolve and filter.
(*b*) Oxalic acid   ...   1 part.
Distilled water   22 „

Mix one part of (*a*) with eight parts of (*b*), and add twelve parts of alcohol.

3. Borax carmine.

Carmine	1 part.
Borax	... 4 ,,
Water	... 56 ,,
Alcohol	... q.s.

Dissolve the borax in the water, and add the carmine; filter, and add to the filtrate two volumes of absolute alcohol.

4.* Ranvier's picro-carmine.

(*a*) Carmine ... ... 1 gramme.
Liquor ammoniæ ... ... 3 c.c.
Water ... ... 10 c.c.

Rub the carmine up with water; add the ammonia, and dissolve with heat carefully applied.

(*b*) Saturated aqueous solution of picric acid, 200 c.c.

Add, when perfectly cold, (*a*) to (*b*), and evaporate with gentle warmth to one-third its bulk. Filter, and keep in stoppered bottle.

5.* Minot's hæmatoxylin dye.

(*a*) Hæmatoxylin crystals ... 3·5 parts.
Absolute alcohol ... ... 100 ,,
Alum ... ... 1 ,,
Water ... ... 300 ,,

Dissolve the hæmatoxylin in alcohol, and add the alum, dissolved in the water.

(*b*) Alum ... ... 1 part.
Water... ... ... 200 ,,

Dissolve. For use, pour a little of (*b*) into a watch-glass, and add sufficient of (*a*) to colour it a light violet tint. Filter prior to staining.

6.* Kleinenberg's hæmatoxylin dye.

(a) Make a saturated solution of crystallized calcium chloride in alcohol (70 per cent.), and add alum to saturation.

(b) Make a saturated solution of alum in alcohol (70 per cent.), and add solution (a) to (b), in the proportion of 1 to 8.

(c) To this resultant mixture, now add a *few drops* of a *barely* alkaline saturated solution of hæmatoxylin.

7.* Aniline blue-black (aqueous solution).

Aniline blue-black ... 1 gramme.
Distilled water ... ... 400 c.c.

Dissolve, filter, and label " 0·25 per cent." Or, make a 1 per cent. solution ; dilute and filter when required.

8. Aniline blue-black (alcoholic solution).

Aniline blue-black ... 1 decigramme.
Water ... ... 4 c.c.
Dissolve and add recti-
fied spirit ... 100 c.c.
Filter, and label " 0·1 per cent."

9. Aniline blue reagent.

Aniline blue ... ... ½ grain.
Distilled water ... ... 1 ounce.
Alcohol ... ... 25 drops.

10. Rutherford's rosanilin dye.

Rosanilin or magenta (crystals) 1 grain.
Absolute alcohol ... ... 100 minims.
Distilled water ... ... 5 ounces.

11. Dyes used in combination.

Aniline blue-black (Aq. sol., 0·25 per cent.) }
          + Picro-carmine.   }

Judson's "Cambridge blue" ⎫
       + Picric acid.   ⎬
Osmic acid (1 per cent.) + Picro-carmine.
Hæmatoxylin   ... + Aniline blue-black
                             (0·5 per cent.).
Eosin (1 to 1,500)   ... + Hæmatoxylin.
Picro-carmine   ... + Iodine green.

12. Alferow's lactate of silver :—

Lactate of silver   ...   ...   1 part.
Distilled water   ...   ...   800 ,,
Concentrated solution of lactic
    acid   ...   .   A few drops.

Note.—" The free acid renders precipitation less easy, and the chloride and albuminate of silver are alone formed."

13. Gerlach's gold and potassium solution.[1]

Double chloride of gold and
    potassium   ...   ...   1 part.
Water, feebly acidulated
    with hydrochloric acid..   10,000 ,,

Sections acquire a pale lilac colour in this solution in from ten to twelve hours.

SOLUTIONS FOR MACERATING AND DISSOCIATING TISSUES.

1. Iodized serum.

Fresh amniotic fluid   ...   100 c.c.
Tincture of iodine   1 c.c.
Carbolic acid   2 drops.

For small quantities, use the aqueous humour from the eye of sheep or ox. Other useful fluids are—

2. Glycerine and water.

---

[1] For staining sections of spinal cord by this method, see Art. " The Spinal Cord," by Gerlach, in Stricker's " Human and Comparative Histology."

3. Chloride of sodium solution (0·75 per cent.).
4. Ammonium and potassium bichromate (1 part to 10,000 of water).
5. Müller's fluid, freely diluted.

MOUNTING MEDIA.

1.* Glycerine (Price's) ; 2.* Glycerine jelly (Rimmington's) ; 3.* Canada balsam ; 4.* Dammar varnish ; 5. Farrant's solution.

3.* Benzole solution of balsam.

Canada balsam evaporated in a water-bath to a brittle, resinous consistency, is dissolved in benzole until it is sufficiently fluid to drop slowly from a glass rod.

4.* Dammar fluid.

Gum dammar, gum mastic,
  of each   ...   . ½ ounce.
Benzole   ...   . 3 „
Dissolve the gum in the benzole and filter.

5.* Farrant's solution.

Glycerine, saturated solution of arsenious acid, gum-arabic, equal parts of each.

Let the mixture stand for several weeks, frequently stirring; and when no more of the gum-arabic is dissolved, filter the solution.

CEMENTS.

1.* Mr. Kitton's cement for glycerine mounting.

White lead, red lead, litharge in powder, equal parts.
Mix well. Keep a stock of the mixture ready ground in a bottle.

For use—rub up a little of the powder with turpentine, and add sufficient gold size to allow it to work readily with a brush. The first coat should always be perfectly dry before the application of a second.

2.* Zinc-white cement.

Gum dammar... ... ... 8 ounces.
Zinc oxide ... . 1 „
Benzole ... ... 8 „

Dissolve the gum in the benzole, add the zinc, and strain through muslin.

3. Marine glue.
4.* Gold size.
5. Brunswick black.
6.* Balsam or dammar varnish.

In addition to the list of reagents above given, the student should supply himself with a bottle of clove oil, glacial acetic acid, and solution of chloral hydrate, 2 per cent.

INDEX.

—◦‹•›—

M

Pardon and Sons, Printers, Paternoster Row, London.

SELECTION

FROM

J. & A. CHURCHILL'S GENERAL CATALOGUE

COMPRISING

ALL RECENT WORKS PUBLISHED BY THEM

ON THE

ART AND SCIENCE OF MEDICINE

N.B.—As far as possible, this List is arranged in the order in which medical study is usually pursued.

J. & A. CHURCHILL publish for the following Institutions and Public Bodies :—

ROYAL COLLEGE OF SURGEONS.
CATALOGUES OF THE MUSEUM.
Twenty-three separate Catalogues (List and Prices can be obtained of J. & A. CHURCHILL).

GUY'S HOSPITAL.
REPORTS BY THE MEDICAL AND SURGICAL STAFF.
Vol. XXVIII., Third Series. 7s. 6d.
FORMULÆ USED IN THE HOSPITAL IN ADDITION TO THOSE
IN THE B.P. 1s. 6d.

LONDON HOSPITAL.
PHARMACOPŒIA OF THE HOSPITAL. 3s.
CLINICAL LECTURES AND REPORTS BY THE MEDICAL AND
SURGICAL STAFF. Vols. I. to IV. 7s. 6d. each.

ST. BARTHOLOMEW'S HOSPITAL.
CATALOGUE OF THE ANATOMICAL AND PATHOLOGICAL
MUSEUM. Vol. I.—Pathology. 15s. Vol. II.—Teratology, Anatomy
and Physiology, Botany. 7s. 6d.

ST. GEORGE'S HOSPITAL.
REPORTS BY THE MEDICAL AND SURGICAL STAFF.
The last Volume (X.) was issued in 1880. Price 7s. 6d.
CATALOGUE OF THE PATHOLOGICAL MUSEUM. 15s.
SUPPLEMENTARY CATALOGUE (1882). 5s.

ST. THOMAS'S HOSPITAL.
REPORTS BY THE MEDICAL AND SURGICAL STAFF.
Annually. Vol. XV., New Series. 7s. 6d.

MIDDLESEX HOSPITAL.
CATALOGUE OF THE PATHOLOGICAL MUSEUM. 12s.

WESTMINSTER HOSPITAL.
REPORTS BY THE MEDICAL AND SURGICAL STAFF.
Annually. Vol. II. 6s.

ROYAL LONDON OPHTHALMIC HOSPITAL.
REPORTS BY THE MEDICAL AND SURGICAL STAFF.
Occasionally. Vol. XI., Part IV. 5s.

OPHTHALMOLOGICAL SOCIETY OF THE UNITED KINGDOM.
TRANSACTIONS.
Vol. VI. 12s. 6d.

MEDICO-PSYCHOLOGICAL ASSOCIATION.
JOURNAL OF MENTAL SCIENCE.
Quarterly. 3s. 6d. each, or 14s. per annum.

PHARMACEUTICAL SOCIETY OF GREAT BRITAIN.
PHARMACEUTICAL JOURNAL AND TRANSACTIONS.
Every Saturday. 4d. each, or 20s. per annum, post free.

BRITISH PHARMACEUTICAL CONFERENCE.
YEAR BOOK OF PHARMACY.
In December. 10s.

BRITISH DENTAL ASSOCIATION.
JOURNAL OF THE ASSOCIATION AND MONTHLY REVIEW
OF DENTAL SURGERY.
On the 15th of each Month. 6d. each, or 7s. per annum, post free.

A SELECTION

FROM

J. & A. CHURCHILL'S GENERAL CATALOGUE,

COMPRISING

ALL RECENT WORKS PUBLISHED BY THEM ON THE ART AND SCIENCE OF MEDICINE.

N.B.—*J. & A. Churchill's Descriptive List of Works on Chemistry, Materia Medica, Pharmacy, Botany, Photography, Zoology, the Microscope, and other Branches of Science, can be had on application.*

Practical Anatomy:
A Manual of Dissections. By CHRISTOPHER HEATH, Surgeon to University College Hospital. Sixth Edition. Revised by RICKMAN J. GODLEE, M.S. Lond., F.R.C.S., Demonstrator of Anatomy in University College, and Assistant Surgeon to the Hospital. Crown 8vo, with 24 Coloured Plates and 274 Engravings, 15s.

Wilson's Anatomist's Vade-Mecum.
Tenth Edition. By GEORGE BUCHANAN, Professor of Clinical Surgery in the University of Glasgow; and HENRY E. CLARK, M.R.C.S., Lecturer on Anatomy at the Glasgow Royal Infirmary School of Medicine. Crown 8vo, with 450 Engravings (including 26 Coloured Plates), 18s.

Braune's Atlas of Topographical Anatomy,
after Plane Sections of Frozen Bodies. Translated by EDWARD BELLAMY, Surgeon to, and Lecturer on Anatomy, &c., at, Charing Cross Hospital. Large Imp. 8vo, with 34 Photo-lithographic Plates and 46 Woodcuts, 40s.

An Atlas of Human Anatomy.
By RICKMAN J. GODLEE, M.S., F.R.C.S., Assistant Surgeon and Senior Demonstrator of Anatomy, University College Hospital. With 48 Imp. 4to Plates (112 figures), and a volume of Explanatory Text. 8vo, £4 14s. 6d.

Harvey's (Wm.) Manuscript
Lectures. Prelectiones Anatomiæ Universalis. Edited, with an Autotype reproduction of the Original, by a Committee of the Royal College of Physicians of London. Crown 4to, half bound in Persian, 52s. 6d.

Anatomy of the Joints of Man.
By HENRY MORRIS, Surgeon to, and Lecturer on Anatomy and Practical Surgery at, the Middlesex Hospital. 8vo, with 44 Lithographic Plates (several being coloured) and 13 Wood Engravings, 16s.

Manual of the Dissection of the Human Body.
By LUTHER HOLDEN, Consulting Surgeon to St. Bartholomew's Hospital. Edited by JOHN LANGTON, F.R.C.S., Surgeon to, and Lecturer on Anatomy at, St. Bartholomew's Hospital. Fifth Edition. 8vo, with 208 Engravings. 20s.

By the same Author.

Human Osteology.
Seventh Edition, edited by CHARLES STEWART, Conservator of the Museum R.C.S., and R. W. REID, M.D., F.R.C.S., Lecturer on Anatomy at St. Thomas's Hospital. 8vo, with 59 Lithographic Plates and 75 Engravings. 16s.

Also.

Landmarks, Medical and Surgical.
Fourth Edition. 8vo. [*In the Press.*

The Student's Guide to Surgical Anatomy.
By EDWARD BELLAMY, F.R.C.S. and Member of the Board of Examiners. Third Edition. Fcap. 8vo, with 81 Engravings. 7s. 6d.

The Anatomical Remembrancer;
or, Complete Pocket Anatomist. Eighth Edition. 32mo, 3s. 6d.

Diagrams of the Nerves of the Human Body,
exhibiting their Origin, Divisions, and Connections, with their Distribution to the Various Regions of the Cutaneous Surface, and to all the Muscles. By W. H. FLOWER, C.B., F.R.S., F.R.C.S. Third Edition, with 6 Plates. Royal 4to, 12s.

General Pathology.
An Introduction to. By JOHN BLAND SUTTON, F.R.C.S., Sir E. Wilson Lecturer on Pathology, R.C.S.; Assistant Surgeon to, and Lecturer on Anatomy at, Middlesex Hospital. 8vo, with 149 Engravings, 14s.

Atlas of Pathological Anatomy.
By Dr. LANCEREAUX. Translated by W. S. GREENFIELD, M.D., Professor of Pathology in the University of Edinburgh. Imp. 8vo, with 70 Coloured Plates, £5 5s.

A Manual of Pathological Anatomy.
By C. HANDFIELD JONES, M.B., F.R.S., and E. H. SIEVEKING, M.D., F.R.C.P. Edited by J. F. PAYNE, M.D., F.R.C.P., Lecturer on General Pathology at St. Thomas's Hospital. Second Edition. Crown 8vo, with 195 Engravings, 16s.

Post-mortem Examinations:
A Description and Explanation of the Method of Performing them, with especial reference to Medico-Legal Practice. By Prof. VIRCHOW. Translated by Dr. T. P. SMITH. Second Edition. Fcap. 8vo, with 4 Plates, 3s. 6d.

The Human Brain:
Histological and Coarse Methods of Research. A Manual for Students and Asylum Medical Officers. By W. BEVAN LEWIS, L.R.C.P. Lond., Medical Superintendent, West Riding Lunatic Asylum. 8vo, with Wood Engravings and Photographs, 8s.

Manual of Physiology:
For the use of Junior Students of Medicine. By GERALD F. YEO, M.D., F.R.C.S., Professor of Physiology in King's College, London. Second Edition. Crown 8vo, with 318 Engravings, 14s.

Principles of Human Physiology.
By W. B. CARPENTER, C.B., M.D., F.R.S. Ninth Edition. By HENRY POWER, M.B., F.R.C.S. 8vo, with 3 Steel Plates and 377 Wood Engravings, 31s. 6d.

Elementary Practical Biology:
Vegetable. By THOMAS W. SHORE, M.D., B.Sc. Lond., Lecturer on Comparative Anatomy at St. Bartholomew's Hospital. 8vo, 6s.

Histology and Histo-Chemistry of Man.
By HEINRICH FREY, Professor of Medicine in Zurich. Translated by ARTHUR E. J. BARKER, Assistant Surgeon to University College Hospital. 8vo, with 608 Engravings, 21s.

A Text-Book of Medical Physics,
for Students and Practitioners. By J. C. DRAPER, M.D., LL.D., Professor of Physics in the University of New York. With 377 Engravings. 8vo, 18s.

Medical Jurisprudence:
Its Principles and Practice. By ALFRED S. TAYLOR, M.D., F.R.C.P., F.R.S. Third Edition, by THOMAS STEVENSON, M.D., F.R.C.P., Lecturer on Medical Jurisprudence at Guy's Hospital. 2 vols. 8vo, with 188 Engravings, 31s. 6d.

By the same Authors,

A Manual of Medical Jurisprudence.
Eleventh Edition. Crown 8vo, with 56 Engravings, 14s.

Also.

Poisons,
In Relation to Medical Jurisprudence and Medicine. Third Edition. Crown 8vo, with 104 Engravings, 16s.

Lectures on Medical Jurisprudence.
By FRANCIS OGSTON, M.D., late Professor in the University of Aberdeen. Edited by FRANCIS OGSTON, Jun., M.D. 8vo, with 12 Copper Plates, 18s.

The Student's Guide to Medical Jurisprudence.
By JOHN ABERCROMBIE, M.D., F.R.C.P., Lecturer on Forensic Medicine to Charing Cross Hospital. Fcap. 8vo, 7s. 6d.

Influence of Sex in Disease.
By W. ROGER WILLIAMS, F.R.C.S., Surgical Registrar to the Middlesex Hospital. 8vo, 3s. 6d.

Microscopical Examination of Drinking Water and of Air.
By J. D. MACDONALD, M.D., F.R.S., Ex-Professor of Naval Hygiene in the Army Medical School. Second Edition. 8vo, with 25 Plates, 7s. 6d.

Pay Hospitals and Paying Wards throughout the World.
By HENRY C. BURDETT. 8vo, 7s.

By the same Author.

Cottage Hospitals — General, Fever, and Convalescent:
Their Progress, Management, and Work. Second Edition, with many Plans and Illustrations. Crown 8vo, 14s.

Hospitals, Infirmaries, and Dispensaries:
Their Construction, Interior Arrangement, and Management; with Descriptions of existing Institutions, and 74 Illustrations. By F. OPPERT, M.D., M.R.C.P.L. Second Edition. Royal 8vo, 12s.

Hospital Construction and Management.
By F. J. MOUAT, M.D., Local Government Board Inspector, and H. SAXON SNELL, Fell. Roy. Inst. Brit. Architects. In 2 Parts, 4to, 15s. each; or, the whole work bound in half calf, with large Map, 54 Lithographic Plates, and 27 Woodcuts, 35s.

Public Health Reports.
By Sir JOHN SIMON, C.B., F.R.S. Edited by EDWARD SEATON, M.D., F.R.C.P. 2 vols. 8vo, with Portrait, 36s.

A Manual of Practical Hygiene.

By F. A. PARKES, M.D., F.R.S. Seventh Edition, by F. DE CHAUMONT, M.D., F.R.S., Professor of Military Hygiene in the Army Medical School. 8vo, with 9 Plates and 100 Engravings, 18s.

A Handbook of Hygiene and Sanitary Science.

By GEO. WILSON, M.A., M.D., F.R.S.E., Medical Officer of Health for Mid-Warwickshire. Sixth Edition. Crown 8vo, with Engravings, 10s. 6d.

By the same Author.

Healthy Life and Healthy Dwellings :

A Guide to Personal and Domestic Hygiene. Fcap. 8vo, 5s.

Sanitary Examinations

Of Water, Air, and Food. A Vade-Mecum for the Medical Officer of Health. By CORNELIUS B. FOX, M.D., F.R.C.P. Second Edition. Crown 8vo, with 110 Engravings, 12s. 6d.

Dangers to Health :

A Pictorial Guide to Domestic Sanitary Defects. By T. PRIDGIN TEALE, M.A., Surgeon to the Leeds General Infirmary. Fourth Edition. 8vo, with 70 Lithograph Plates (mostly coloured), 10s.

Manual of Anthropometry :

A Guide to the Measurement of the Human Body, containing an Anthropometrical Chart and Register, a Systematic Table of Measurements, &c. By CHARLES ROBERTS, F.R.C.S. 8vo, with numerous Illustrations and Tables, 8s. 6d.

By the same Author.

Detection of Colour-Blindness and Imperfect Eyesight.

8vo, with a Table of Coloured Wools, and Sheet of Test-types, 5s.

Illustrations of the Influence of the Mind upon the Body in Health and Disease :

Designed to elucidate the Action of the Imagination. By DANIEL HACK TUKE, M.D., F.R.C.P., LL.D. Second Edition. 2 vols. crown 8vo, 15s.

By the same Author.

Sleep-Walking and Hypnotism.

8vo, 5s.

A Manual of Psychological Medicine.

With an Appendix of Cases. By JOHN C. BUCKNILL, M.D., F.R.S., and D. HACK TUKE, M.D., F.R.C.P. Fourth Edition. 8vo, with 12 Plates (30 Figures) and Engravings, 25s.

Mental Affections of Childhood and Youth

(Lettsomian Lectures for 1887, &c.). By J. LANGDON DOWN, M.D., F.R.C.P., Senior Physician to the London Hospital. 8vo, 6s.

Private Treatment of the Insane

as Single Patients. By EDWARD EAST, M.R.C.S., L.S.A. Crown 8vo, 2s. 6d.

Mental Diseases.

Clinical Lectures. By T. S. CLOUSTON, M.D., F.R.C.P. Edin., Lecturer on Mental Diseases in the University of Edinburgh. Second Edition. Crown 8vo, with 8 Plates (6 Coloured), 12s. 6d.

Manual of Midwifery.

By ALFRED L. GALABIN, M.A., M.D., F.R.C.P., Obstetric Physician to, and Lecturer on Midwifery, &c. at, Guy's Hospital. Crown 8vo, with 227 Engravings, 15s.

The Student's Guide to the Practice of Midwifery.

By D. LLOYD ROBERTS, M.D., F.R.C.P., Lecturer on Clinical Midwifery and Diseases of Women at the Owens College; Obstetric Physician to the Manchester Royal Infirmary. Third Edition. Fcap. 8vo, with 2 Coloured Plates and 127 Wood Engravings, 7s. 6d.

Lectures on Obstetric Operations :

Including the Treatment of Hæmorrhage, and forming a Guide to the Management of Difficult Labour. By ROBERT BARNES, M.D., F.R.C.P., Consulting Obstetric Physician to St. George's Hospital. Fourth Edition. 8vo, with 121 Engravings, 12s. 6d.

By the same Author.

A Clinical History of Medical and Surgical Diseases of Women.

Second Edition. 8vo, with 181 Engravings, 28s.

Clinical Lectures on Diseases of Women :

Delivered in St. Bartholomew's Hospital, by J. MATTHEWS DUNCAN, M.D., LL.D., F.R.S. Third Edition. 8vo, 16s.

By the same Author.

Sterility in Woman.

Being the Gulstonian Lectures, delivered in the Royal College of Physicians, in Feb., 1883. 8vo, 6s.

Notes on Diseases of Women :

Specially designed to assist the Student in preparing for Examination. By J. J. REYNOLDS, L.R.C.P., M.R.C.S. Third Edition. Fcap. 8vo, 2s. 6d.

By the same Author.

Notes on Midwifery :

Specially designed for Students preparing for Examination. Second Edition. Fcap. 8vo, with 15 Engravings, 4s.

Dysmenorrhœa, its Pathology and Treatment.

By HEYWOOD SMITH, M.D. Crown 8vo, with Engravings, 4s. 6d.

A Manual of Obstetrics.

By A. F. A. KING, A.M., M.D., Professor of Obstetrics, &c., in the Columbian University, Washington, and the University of Vermont. Third Edition. Crown 8vo, with 102 Engravings, 8s.

The Student's Guide to the Diseases of Women. By ALFRED L. GALABIN, M.D., F.R.C.P., Obstetric Physician to Guy's Hospital. Fourth Edition. Fcap. 8vo, with 94 Engravings, 7s. 6d.

West on the Diseases of Women. Fourth Edition, revised by the Author, with numerous Additions by J. MATTHEWS DUNCAN, M.D., F.R.C.P., F.R.S.E., Obstetric Physician to St. Bartholomew's Hospital. 8vo, 16s.

Obstetric Aphorisms: For the Use of Students commencing Midwifery Practice. By JOSEPH G. SWAYNE, M.D. Eighth Edition. Fcap. 8vo, with Engravings, 3s. 6d.

Handbook of Midwifery for Midwives: By J. E. BURTON, L.R.C.P. Lond., Surgeon to the Hospital for Women, Liverpool. Second Edition. With Engravings. Fcap. 8vo, 6s.

A Handbook of Uterine Therapeutics, and of Diseases of Women. By E. J. TILT, M.D., M.R.C.P, Fourth Edition. Post 8vo, 10s.

By the same Author.

The Change of Life In Health and Disease: A Clinical Treatise on the Diseases of the Nervous System incidental to Women at the Decline of Life. Fourth Edition. 8vo, 10s. 6d.

Diseases of the Uterus, Ovaries, and Fallopian Tubes: A Practical Treatise by A. COURTY, Professor of Clinical Surgery, Montpellier. Translated from Third Edition by his Pupil, AGNES McLAREN, M.D., M.K.Q.C.P.I., with Preface by J. MATTHEWS DUNCAN, M.D., F.R.C.P. 8vo, with 424 Engravings, 24s.

The Female Pelvic Organs: Their Surgery, Surgical Pathology, and Surgical Anatomy. In a Series of Coloured Plates taken from Nature; with Commentaries, Notes, and Cases. By HENRY SAVAGE, M.D., F.R.C.S., Consulting Officer of the Samaritan Free Hospital. Fifth Edition. Roy. 4to, with 17 Lithographic Plates (15 coloured) and 52 Woodcuts, £1 15s.

A Practical Treatise on the Diseases of Women. By T. GAILLARD THOMAS, M.D., Professor of Diseases of Women in the College of Physicians and Surgeons, New York. Fifth Edition. Roy. 8vo, with 266 Engravings, 25s.

Backward Displacements of the Uterus and Prolapsus Uteri: Treatment by the New Method of Shortening the Round Ligaments. By WILLIAM ALEXANDER, M.D., M.Ch.Q.U.I., F.R.C.S., Surgeon to the Liverpool Infirmary. Crown 8vo, with Engravings, 3s. 6d.

Gynæcological Operations: (Handbook of). By ALBAN II.G. DORAN, F.R.C.S., Surgeon to the Samaritan Hospital. 8vo, with 167 Engravings, 15s.

Abdominal Surgery. By J. GREIG SMITH, M.A., F.R.S.E., Surgeon to the Bristol Royal Infirmary. 8vo, with 43 Engravings, 15s.

Ovarian and Uterine Tumours: Their Pathology and Surgical Treatment. By Sir T. SPENCER WELLS, Bart., F.R.C.S., Consulting Surgeon to the Samaritan Hospital. 8vo, with Engravings, 21s.

By the same Author.

Abdominal Tumours: Their Diagnosis and Surgical Treatment. 8vo, with Engravings, 3s. 6d.

The Student's Guide to Diseases of Children. By JAS. F. GOODHART, M.D., F.R.C.P., Physician to Guy's Hospital, and to the Evelina Hospital for Sick Children. Second Edition. Fcap. 8vo, 10s. 6d.

Diseases of Children. For Practitioners and Students. By W. II. DAY, M.D., Physician to the Samaritan Hospital. Second Edition. Crown 8vo, 12s. 6d.

A Practical Treatise on Disease in Children. By EUSTACE SMITH, M.D., Physician to the King of the Belgians, Physician to the East London Hospital for Children. 8vo, 22s.

By the same Author.

Clinical Studies of Disease in Children. Second Edition. Post 8vo, 7s. 6d.

Also.

The Wasting Diseases of Infants and Children. Fourth Edition. Post 8vo, 8s. 6d.

A Practical Manual of the Diseases of Children. With a Formulary. By EDWARD ELLIS, M.D. Fifth Edition. Crown 8vo, 10s.

A Manual for Hospital Nurses and others engaged in Attending on the Sick. By EDWARD J. DOMVILLE, Surgeon to the Exeter Lying-in Charity. Fifth Edition. Crown 8vo, 2s. 6d.

A Manual of Nursing, Medical and Surgical. By CHARLES J. CULLINGWORTH, M.D., Physician to St. Mary's Hospital, Manchester. Second Edition. Fcap. 8vo, with Engravings, 3s. 6d.

By the same Author.

A Short Manual for Monthly Nurses. Second Edition. Fcap. 8vo, 1s. 6d.

Diseases and their Commencement. Lectures to Trained Nurses. By DONALD W. C. HOOD, M.D., M.R.C.P., Physician to the West London Hospital. Crown 8vo, 2s. 6d.

Notes on Fever Nursing.
By J. W. ALLAN, M.B., Physician, Superintendent Glasgow Fever Hospital. Crown 8vo, with Engravings, 2s. 6d.

By the same Author.

Outlines of Infectious Diseases :
For the use of Clinical Students. Fcap. 8vo.

Hospital Sisters and their Duties.
By EVA C. E. LÜCKES, Matron to the London Hospital. Crown 8vo, 2s. 6d.

Infant Feeding and its Influence on Life ;
By C. H. F. ROUTH, M.D., Physician to the Samaritan Hospital. Fourth Edition. Fcap. 8vo. [*Preparing.*

Manual of Botany:
Including the Structure, Classification, Properties, Uses, and Functions of Plants. By ROBERT BENTLEY, Professor of Botany in King's College and to the Pharmaceutical Society. Fifth Edition. Crown 8vo, with 1,178 Engravings, 15s.

By the same Author.

The Student's Guide to Structural, Morphological, and Physiological Botany.
With 660 Engravings. Fcap. 8vo, 7s. 6d.

Also.

The Student's Guide to Systematic Botany,
including the Classification of Plants and Descriptive Botany. Fcap. 8vo, with 350 Engravings, 3s. 6d.

Medicinal Plants :
Being descriptions, with original figures, of the Principal Plants employed in Medicine, and an account of their Properties and Uses. By Prof. BENTLEY and Dr. H. TRIMEN. In 4 vols., large 8vo, with 306 Coloured Plates, bound in Half Morocco, Gilt Edges, £11 11s.

The National Dispensatory :
Containing the Natural History, Chemistry, Pharmacy, Actions and Uses of Medicines. By ALFRED STILLÉ, M.D., LL.D., and JOHN M. MAISCH, Ph.D. Fourth Edition. 8vo, with 311 Engravings, 36s.

Royle's Manual of Materia Medica and Therapeutics.
Sixth Edition, including additions and alterations in the B.P. 1885. By JOHN HARLEY, M.D., Physician to St. Thomas's Hospital. Crown 8vo, with 139 Engravings, 15s.

Materia Medica and Therapeutics :
Vegetable Kingdom — Organic Compounds — Animal Kingdom. By CHARLES D. F. PHILLIPS, M.D., F.R.S. Edin., late Lecturer on Materia Medica and Therapeutics at the Westminster Hospital Medical School. 8vo, 25s.

The Student's Guide to Materia Medica and Therapeutics.
By JOHN C. THOROWGOOD, M.D., F.R.C.P. Second Edition. Fcap. 8vo, 7s.

Materia Medica.
A Manual for the use of Students. By ISAMBARD OWEN, M.D., F.R.C.P., Lecturer on Materia Medica, &c., to St. George's Hospital. Second Edition. Crown 8vo, 6s. 6d.

The Pharmacopœia of the London Hospital.
Compiled under the direction of a Committee appointed by the Hospital Medical Council. Fcap. 8vo, 3s.

A Companion to the British Pharmacopœia.
By PETER SQUIRE, Revised by his Sons, P. W. and A. H. SQUIRE. 14th Edition. 8vo, 10s. 6d.

By the same Authors.

The Pharmacopœias of the London Hospitals,
arranged in Groups for Easy Reference and Comparison. Fifth Edition. 18mo, 6s.

The Prescriber's Pharmacopœia:
The Medicines arranged in Classes according to their Action, with their Composition and Doses. By NESTOR J. C. TIRARD, M.D., F.R.C.P., Professor of Materia Medica and Therapeutics in King's College, London. Sixth Edition. 32mo, bound in leather, 3s.

A Treatise on the Principles and Practice of Medicine.
Sixth Edition. By AUSTIN FLINT, M.D., W.H. WELCH, M.D., and AUSTIN FLINT, jun., M.D. 8vo, with Engravings, 26s.

Climate and Fevers of India,
with a series of Cases (Croonian Lectures, 1882). By Sir JOSEPH FAYRER, K.C.S.I., M.D. 8vo, with 17 Temperature Charts, 12s.

Family Medicine for India.
A Manual. By WILLIAM J. MOORE, M.D., C.I.E., Honorary Surgeon to the Viceroy of India. Published under the Authority of the Government of India. Fifth Edition. Post 8vo, with Engravings. [*In the Press.*

By the same Author.

A Manual of the Diseases of India :
With a Compendium of Diseases generally. Second Edition. Post 8vo, 10s.

Also.

Health-Resorts for Tropical Invalids,
in India, at Home, and Abroad. Post 8vo, 5s.

Practical Therapeutics :
A Manual. By EDWARD J. WARING, C.I.E., M.D., F.R.C.P., and DUDLEY W. BUXTON, M.D., B.S. Lond. Fourth Edition. Crown 8vo, 14s.

By the same Author.

Bazaar Medicines of India,
And Common Medical Plants : With Full Index of Diseases, indicating their Treatment by these and other Agents procurable throughout India, &c. Fourth Edition. Fcap. 8vo, 5s.

A Commentary on the Diseases of India.
By NORMAN CHEVERS, C.I.E., M.D., F.R.C.S., Deputy Surgeon-General H.M. Indian Army. 8vo, 24s.

The Principles and Practice of Medicine.
By C. HILTON FAGGE, M.D. Edited by P. H. PYE-SMITH, M.D., F.R.C.P., Physician to, and Lecturer on Medicine at, Guy's Hospital. 2 vols. 8vo, 1860 pp. Cloth, 36s. ; Half Persian, 42s.

The Student's Guide to the Practice of Medicine.
By MATTHEW CHARTERIS, M.D., Professor of Materia Medica in the University of Glasgow. Fourth Edition. Fcap. 8vo, with Engravings on Copper and Wood. 9s.

Hooper's Physicians' Vade-Mecum.
A Manual of the Principles and Practice of Physic. Tenth Edition. By W. A. GUY, F.R.C.P., F.R.S., and J. HARLEY, M.D., F.R.C.P. With 118 Engravings. Fcap. 8vo, 12s. 6d.

The Student's Guide to Clinical Medicine and Case-Taking.
By FRANCIS WARNER, M.D., F.R.C.P., Physician to the London Hospital. Second Edition. Fcap. 8vo, 5s.

How to Examine the Chest :
Being a Practical Guide for the use of Students. By SAMUEL WEST, M.D., F.R.C.P., Physician to the City of London Hospital for Diseases of the Chest; Assistant Physician to St. Bartholomew's Hospital. With 42 Engravings. Fcap. 8vo, 5s.

The Contagiousness of Pulmonary Consumption, and its Antiseptic Treatment.
By J. BURNEY YEO, M.D., Physician to King's College Hospital. Crown 8vo, 3s. 6d.

The Operative Treatment of Intra-thoracic Effusion.
Fothergillian Prize Essay. By NORMAN PORRITT, L.R.C.P. Lond., M.R.C.S. With Engravings. Crown 8vo, 6s.

Diseases of the Chest :
Contributions to their Clinical History, Pathology, and Treatment. By A. T. HOUGHTON WATERS, M.D., Physician to the Liverpool Royal Infirmary. Second Edition. 8vo, with Plates, 15s.

Pulmonary Consumption :
A Practical Treatise on its Cure with Medicinal, Dietetic, and Hygienic Remedies. By JAMES WEAVER, M.D., L.R.C.P. Crown 8vo, 2s.

Croonian Lectures on Some Points in the Pathology and Treatment of Typhoid Fever.
By WILLIAM CAYLEY, M.D., F.R.C.P., Physician to the Middlesex and the London Fever Hospitals. Crown 8vo, 4s. 6d.

The Student's Guide to Medical Diagnosis.
By SAMUEL FENWICK, M.D., F.R.C.P., Physician to the London Hospital, and BEDFORD FENWICK, M.D., M.R.C.P. Sixth Edition. Fcap. 8vo, with 114 Engravings, 7s.

By the same Author.

The Student's Outlines of Medical Treatment.
Second Edition. Fcap. 8vo, 7s.

Also.

On Chronic Atrophy of the Stomach, and on the Nervous Affections of the Digestive Organs.
8vo, 8s.

The Microscope in Medicine.
By LIONEL S. BEALE, M.B., F.R.S., Physician to King's College Hospital. Fourth Edition. 8vo, with 86 Plates, 21s.

Also.

On Slight Ailments :
Their Nature and Treatment. Second Edition. 8vo, 5s.

Medical Lectures and Essays.
By GEORGE JOHNSON, M.D., F.R.C.P., F.R.S., Consulting Physician to King's College Hospital. 8vo.

The Spectroscope in Medicine.
By CHARLES A. MACMUNN, B.A., M.D. 8vo, with 3 Chromo-lithographic Plates of Physiological and Pathological Spectra, and 13 Engravings, 9s.

Notes on Asthma :
Its Forms and Treatment. By JOHN C. THOROWGOOD, M.D., Physician to the Hospital for Diseases of the Chest. Third Edition. Crown 8vo, 4s. 6d.

What is Consumption ?
By G. W. HAMBLETON, L.K.Q.C.P.I. Crown 8vo, 2s. 6d.

Winter Cough
(Catarrh, Bronchitis, Emphysema, Asthma). By HORACE DOBELL, M.D., Consulting Physician to the Royal Hospital for Diseases of the Chest. Third Edition. 8vo, with Coloured Plates, 10s. 6d.

By the same Author.

Loss of Weight, Blood-Spitting, and Lung Disease.
Second Edition. 8vo, with Chromo-lithograph, 10s. 6d.

Also.

The Mont Dore Cure, and the Proper Way to Use it.
8vo, 7s. 6d.

Vaccinia and Variola :
A Study of their Life History. By JOHN B. BUIST, M.D., F.R.S.E., Teacher of Vaccination for the Local Government Board. Crown 8vo, with 24 Coloured Plates, 7s. 6d.

Treatment of Some of the Forms of Valvular Disease of the Heart.
By A. E. SANSOM, M.D., F.R.C.P., Physician to the London Hospital. Second Edition. Fcap. 8vo, with 26 Engravings, 4s. 6d.

Diseases of the Heart and Aorta :
Clinical Lectures. By G. W. BALFOUR, M.D., F.R.C.P., F.R.S. Edin., late Senior Physician and Lecturer on Clinical Medicine, Royal Infirmary, Edinburgh. Second Edition. 8vo, with Chromo-lithograph and Wood Engravings, 12s. 6d.

Medical Ophthalmoscopy :
A Manual and Atlas. By WILLIAM R. GOWERS, M.D., F.R.C.P., Professor of Clinical Medicine in University College, and Physician to the Hospital. Second Edition, with Coloured Autotype and Lithographic Plates and Woodcuts. 8vo, 18s.

By the same Author.

Pseudo-Hypertrophic Muscular Paralysis : A Clinical Lecture. 8vo, with Engravings and Plate, 3s. 6d.

Also.

Diagnosis of Diseases of the Spinal Cord. Third Edition. 8vo, with Engravings, 4s. 6d.

Also.

Diagnosis of Diseases of the Brain. Second Edition. 8vo, with Engravings, 7s. 6d.

Also.

A Manual of Diseases of the Nervous System. Vol. I. Diseases of the Spinal Cord and Nerves. Roy. 8vo, with 171 Engravings (many figures), 12s.6d.

Diseases of the Nervous System.
Lectures delivered at Guy's Hospital. By SAMUEL WILKS, M.D., F.R.S. Second Edition. 8vo, 18s.

Diseases of the Nervous System:
Especially in Women. By S. WEIR MITCHELL, M.D., Physician to the Philadelphia Infirmary for Diseases of the Nervous System. Second Edition. 8vo, with 5 Plates, 8s.

Nerve Vibration and Excitation,
as Agents in the Treatment of Functional Disorder and Organic Disease. By J. MORTIMER GRANVILLE, M.D. 8vo, 5s.

By the same Author.

Gout in its Clinical Aspects.
Crown 8vo, 6s.

Regimen to be adopted in Cases
of Gout. By WILHELM EBSTEIN, M.D., Professor of Clinical Medicine in Göttingen. Translated by JOHN SCOTT, M.A., M.B. 8vo, 2s. 6d.

Diseases of the Nervous System.
Clinical Lectures. By THOMAS BUZZARD, M.D., F.R.C.P., Physician to the National Hospital for the Paralysed and Epileptic. With Engravings, 8vo. 15s.

By the same Author.

Some Forms of Paralysis from
Peripheral Neuritis : of Gouty, Alcoholic, Diphtheritic, and other origin. Crown 8vo, 5s.

Diseases of the Liver :
With and without Jaundice. By GEORGE HARLEY, M.D., F.R.C.P., F.R.S. 8vo, with 2 Plates and 36 Engravings, 21s.

By the same Author.

Inflammations of the Liver, and
their Sequelæ. Crown 8vo, with Engravings, 5s.

Gout, Rheumatism,
And the Allied Affections ; with Chapters on Longevity and Sleep. By PETER HOOD, M.D. Third Edition. Crown 8vo, 7s. 6d.

Diseases of the Stomach :
The Varieties of Dyspepsia, their Diagnosis and Treatment. By S. O. HABERSHON, M.D., F.R.C.P. Third Edition. Crown 8vo, 5s.

By the same Author.

Pathology of the Pneumogastric Nerve : Lumleian Lectures for 1876. Second Edition. Post 8vo, 4s.

Also.

Diseases of the Abdomen,
Comprising those of the Stomach and other parts of the Alimentary Canal, Œsophagus, Cæcum, Intestines, and Peritoneum. Third Edition. 8vo, with 5 Plates, 21s.

Also.

Diseases of the Liver,
Their Pathology and Treatment. Lettsomian Lectures. Second Edition. Post 8vo, 4s.

Acute Intestinal Strangulation,
And Chronic Intestinal Obstruction (Mode of Death from). By THOMAS BRYANT, F.R.C.S., Senior Surgeon to Guy's Hospital. 8vo, 3s.

A Treatise on the Diseases of
the Nervous System. By JAMES ROSS, M.D., F.R.C.P., Assistant Physician to the Manchester Royal Infirmary. Second Edition. 2 vols. 8vo, with Lithographs, Photographs, and 332 Woodcuts, 52s. 6d.

By the same Author.

Handbook of the Diseases of
the Nervous System. Roy. 8vo, with 184 Engravings, 18s.

Also.

Aphasia :
Being a Contribution to the Subject of the Dissolution of Speech from Cerebral Disease. 8vo, with Engravings, 4s. 6d.

Spasm in Chronic Nerve Disease.
(Gulstonian Lectures.) By SEYMOUR J. SHARKEY, M.A., M.B., F.R.C.P., Assistant Physician to, and Joint Lecturer on Pathology at, St. Thomas's Hospital. 8vo, with Engravings, 5s.

On Megrim, Sick Headache, and
some Allied Disorders : A Contribution to the Pathology of Nerve Storms. By E. LIVEING, M.D., F.R.C.P. 8vo, 15s.

Food and Dietetics,

Physiologically and Therapeutically Considered. By F. W. PAVY, M.D., F.R.S., Physician to Guy's Hospital. Second Edition. 8vo, 15s.

By the same Author.

Croonian Lectures on Certain Points connected with Diabetes. 8vo, 4s. 6d.

Headaches :

Their Nature, Causes, and Treatment. By W. H. DAY, M.D., Physician to the Samaritan Hospital. Fourth Edition. Crown 8vo, with Engravings. [*In the Press.*

Health Resorts at Home and Abroad. By MATTHEW CHARTERIS, M.D., Physician to the Glasgow Royal Infirmary. Second Edition. Crown 8vo, with Map, 5s. 6d.

The Principal Southern and Swiss Health-Resorts : their Climate and Medical Aspect. By WILLIAM MARCET, M.D., F.R.C.P., F.R.S. With Illustrations. Crown 8vo, 7s. 6d.

Winter and Spring

On the Shores of the Mediterranean. By HENRY BENNET, M.D. Fifth Edition. Post 8vo, with numerous Plates, Maps, and Engravings, 12s. 6d.

By the same Author.

Treatment of Pulmonary Consumption by Hygiene, Climate, and Medicine. Third Edition. 8vo, 7s. 6d.

Medical Guide to the Mineral Waters of France and its Wintering Stations. With a Special Map. By A. VINTRAS, M.D., Physician to the French Embassy, and to the French Hospital, London. Crown 8vo, 8s.

The Ocean as a Health-Resort :

A Practical Handbook of the Sea, for the use of Tourists and Health-Seekers. By WILLIAM S. WILSON, L.R.C.P. Second Edition, with Chart of Ocean Routes, &c. Crown 8vo, 7s. 6d.

Ambulance Handbook for Volunteers and Others. By J. ARDAVON RAYE, L.K. & Q.C.P.I., L.R.C.S.I., late Surgeon to H.B.M. Transport No. 14, Zulu Campaign, and Surgeon E.I.R. Rifles. 8vo, with 16 Plates (50 figures), 3s. 6d.

Ambulance Lectures :

To which is added a NURSING LECTURE. By JOHN M. H. MARTIN, Honorary Surgeon to the Blackburn Infirmary. Crown 8vo, with 53 Engravings, 2s.

Commoner Diseases and Accidents to Life and Limb: their Prevention and Immediate Treatment. By M. M. BASIL, M.A., M.B., C.M. Crown 8vo, 2s. 6d.

Handbook of Medical and Surgical Electricity. By HERBERT TIBBITS, M.D., F.R.C.P.E., Senior Physician to the West London Hospital for Paralysis and Epilepsy. Second Edition. 8vo, with 95 Engravings, 9s.

By the same Author.

How to Use a Galvanic Battery in Medicine and Surgery. Third Edition. 8vo, with Engravings, 4s.

Also.

A Map of Ziemssen's Motor Points of the Human Body : A Guide to Localised Electrisation. Mounted on Rollers, 35 × 21. With 20 Illustrations, 5s.   *Also.*

Electrical and Anatomical Demonstrations. A Handbook for Trained Nurses and Masseuses. Crown 8vo, with 44 Illustrations, 5s.

Spina Bifida :

Its Treatment by a New Method. By JAS. MORTON, M.D., L.R.C.S.E., Professor of Materia Medica in Anderson's College, Glasgow. 8vo, with Plates, 7s. 6d.

Surgical Emergencies :

Together with the Emergencies attendant on Parturition and the Treatment of Poisoning. By W. PAUL SWAIN, F.R.C.S., Surgeon to the South Devon and East Cornwall Hospital. Fourth Edition. Crown 8vo, with 120 Engravings, 5s.

Operative Surgery in the Calcutta Medical College Hospital. Statistics, Cases, and Comments. By KENNETH McLEOD, A.M., M.D., F.R.C.S.E., Surgeon-Major, Indian Medical Service, Professor of Surgery in Calcutta Medical College. 8vo, with Illustrations. 12s. 6d.

Surgical Pathology and Morbid Anatomy (Student's Guide). By ANTHONY A. BOWLBY, F.R.C.S., Surgical Registrar and Demonstrator of Surgical Pathology to St. Bartholomew's Hospital. Fcap. 8vo, with 135 Engravings, 9s.

A Course of Operative Surgery.

By CHRISTOPHER HEATH, Surgeon to University College Hospital. Second Edition. With 20 coloured Plates (180 figures) from Nature, by M. LÉVEILLÉ, and several Woodcuts. Large 8vo, 30s.

By the same Author.

The Student's Guide to Surgical Diagnosis. Second Edition. Fcap. 8vo, 6s. 6d.   *Also.*

Manual of Minor Surgery and Bandaging. For the use of House-Surgeons, Dressers, and Junior Practitioners. Eighth Edition. Fcap. 8vo, with 142 Engravings, 6s.   *Also.*

Injuries and Diseases of the Jaws. Third Edition. 8vo, with Plate and 206 Wood Engravings, 14s.

The Practice of Surgery:
A Manual. By THOMAS BRYANT, Surgeon to Guy's Hospital. Fourth Edition. 2 vols. crown 8vo, with 750 Engravings (many being coloured), and including 6 chromo plates, 32s.

Surgery: its Theory and Practice (Student's Guide). By WILLIAM J. WALSHAM, F.R.C.S., Assistant Surgeon to St. Bartholomew's Hospital. Fcap. 8vo, with 236 Engravings, 10s. 6d.

The Surgeon's Vade-Mecum:
A Manual of Modern Surgery. By R. DRUITT, F.R.C.S. Twelfth Edition. By STANLEY BOYD, M.B., F.R.C.S. Assistant Surgeon and Pathologist to Charing Cross Hospital. Crown 8vo, with 373 Engravings 16s.

Regional Surgery:
Including Surgical Diagnosis. A Manual for the use of Students. By F. A. SOUTHAM, M.A., M.B., F.R.C.S., Assistant Surgeon to the Manchester Royal Infirmary. Part I. The Head and Neck. Crown 8vo, 6s. 6d. — Part II. The Upper Extremity and Thorax. Crown 8vo, 7s. 6d. Part III. The Abdomen and Lower Extremity. Crown 8vo, 7s.

Illustrations of Clinical Surgery.
By JONATHAN HUTCHINSON, F.R.S., Senior Surgeon to the London Hospital. In occasional fasciculi. I. to XIX., 6s. 6d. each. Fasciculi I. to X. bound, with Appendix and Index, £3 10s.

By the same Author.

Pedigree of Disease:
Being Six Lectures on Temperament, Idiosyncrasy, and Diathesis. 8vo, 5s.

Treatment of Wounds and Fractures. Clinical Lectures. By SAMPSON GAMGEE, F.R.S.E., Surgeon to the Queen's Hospital, Birmingham. Second Edition. 8vo, with 40 Engravings, 10s.

Electricity and its Manner of Working in the Treatment of Disease. By WM. E. STEAVENSON, M.D., Physician and Electrician to St. Bartholomew's Hospital. 8vo, 4s. 6d.

Lectures on Orthopædic Surgery. By BERNARD E. BRODHURST, F.R.C.S., Surgeon to the Royal Orthopædic Hospital. Second Edition. 8vo, with Engravings, 12s. 6d.

By the same Author.

On Anchylosis, and the Treatment for the Removal of Deformity and the Restoration of Mobility in Various Joints. Fourth Edition. 8vo, with Engravings, 5s.

Also.

Curvatures and Diseases of the Spine. Third Edition. 8vo, with Engravings, 6s.

Diseases of Bones and Joints.
By CHARLES MACNAMARA, F.R.C.S., Surgeon to, and Lecturer on Surgery at, the Westminster Hospital. 8vo, with Plates and Engravings, 12s.

Injuries of the Spine and Spinal Cord, and NERVOUS SHOCK, in their Surgical and Medico-Legal Aspects. By HERBERT W. PAGE, M.C. Cantab., F.R.C.S., Surgeon to St. Mary's Hospital. Second Edition, post 8vo, 10s.

Face and Foot Deformities.
By FREDERICK CHURCHILL, C.M., Surgeon to the Victoria Hospital for Children. 8vo, with Plates and Illustrations, 10s. 6d.

Clubfoot:
Its Causes, Pathology, and Treatment. By WM. ADAMS, F.R.C.S., Surgeon to the Great Northern Hospital. Second Edition. 8vo, with 106 Engravings and 6 Lithographic Plates, 15s.

By the same Author.

On Contraction of the Fingers, and its Treatment by Subcutaneous Operation; and on Obliteration of Depressed Cicatrices, by the same Method. 8vo, with 30 Engravings, 4s. 6d.

Also.

Lateral and other Forms of Curvature of the Spine: Their Pathology and Treatment. Second Edition. 8vo, with 5 Lithographic Plates and 72 Wood Engravings, 10s. 6d.

Spinal Curvatures:
Treatment by Extension and Jacket; with Remarks on some Affections of the Hip, Knee, and Ankle-joints. By H. MACNAUGHTON JONES, M.D., F.R.C.S. I. and Edin. Post 8vo, with 63 Engravings, 4s. 6d.

On Diseases and Injuries of the Eye: A Course of Systematic and Clinical Lectures to Students and Medical Practitioners. By J. R. WOLFE, M.D., F.R.C.S.E., Lecturer on Ophthalmic Medicine and Surgery in Anderson's College, Glasgow. With 10 Coloured Plates and 157 Wood Engravings. 8vo, £1 1s.

Hints on Ophthalmic Out-Patient Practice. By CHARLES HIGGENS, Ophthalmic Surgeon to Guy's Hospital. Third Edition. Fcap. 8vo, 3s.

Short Sight, Long Sight, and Astigmatism. By GEORGE F. HELM, M.A., M.D., F.R.C.S., formerly Demonstrator of Anatomy in the Cambridge Medical School. Crown 8vo, with 35 Engravings, 3s. 6d.

Manual of the Diseases of the Eye. By CHARLES MACNAMARA, F.R.C.S., Surgeon to Westminster Hospital. Fourth Edition. Crown 8vo, with 4 Coloured Plates and 66 Engravings, 10s. 6d.

The Student's Guide to Diseases
of the Eye. By EDWARD NETTLESHIP,
F.R.C.S., Ophthalmic Surgeon to St.
Thomas's Hospital. Fourth Edition.
Fcap. 8vo, with 164 Engravings and a
Set of Coloured Papers illustrating Colour-
Blindness, 7s. 6d.

Normal and Pathological Histology of the Human Eye and Eyelids. By C. FRED. POLLOCK,
M.D., F.R.C.S. and F.R.S.E., Surgeon
for Diseases of the Eye to Anderson's
College Dispensary, Glasgow. Crown
8vo, with 100 Plates (230 drawings), 15s.

Atlas of Ophthalmoscopy.
Composed of 12 Chromo - lithographic
Plates (59 Figures drawn from nature)
and Explanatory Text. By RICHARD
LIEBREICH, M.R.C.S. Translated by H.
ROSBOROUGH SWANZY, M.B. Third
edition, 4to, 40s.

Glaucoma :
Its Causes, Symptoms, Pathology, and
Treatment. By PRIESTLEY SMITH,
M.R.C.S., Ophthalmic Surgeon to the
Queen's Hospital, Birmingham. 8vo,
with Lithographic Plates, 10s. 6d.

Refraction of the Eye :
A Manual for Students. By GUSTAVUS
HARTRIDGE, F.R.C.S., Assistant Physician to the Royal Westminster Ophthalmic
Hospital. Second Edition. Crown 8vo,
with Lithographic Plate and 94 Woodcuts,
5s. 6d.

Squint :
(Clinical Investigations on). By C.
SCHWEIGGER, M.D., Professor of Oph-
thalmology in the University of Berlin.
Edited by GUSTAVUS HARTRIDGE,
F.R.C.S. 8vo, 5s.

The Electro-Magnet,
And its Employment in Ophthalmic Surgery. By SIMEON SNELL, Ophthalmic
Surgeon to the Sheffield General Infirmary, &c. Crown 8vo, 3s. 6d.

Practitioner's Handbook of
Diseases of the Ear and Naso-
Pharynx. By H. MACNAUGHTON
JONES, M.D., late Professor of the Queen's
University in Ireland, Surgeon to the Cork
Ophthalmic and Aural Hospital. Third
Edition of "Aural Surgery." Roy. 8vo,
with 128 Engravings, 6s.

By the same Author.

Atlas of Diseases of the Membrana Tympani. In Coloured
Plates, containing 62 Figures, with Text.
Crown 4to, 21s.

Endemic Goitre or Thyreocele :
Its Etiology, Clinical Characters, Patho-
logy, Distribution, Relations to Cretinism,
Myxœdema, &c., and Treatment. By
WILLIAM ROBINSON, M.D. 8vo, 5s.

Diseases and Injuries of the
Ear. By Sir WILLIAM B. DALBY, Aural
Surgeon to St. George's Hospital. Third
Edition. Crown 8vo, with Engravings,
7s. 6d.

By the Same Author.

Short Contributions to Aural
Surgery, between 1875 and 1886.
8vo, with Engravings, 3s. 6d.

Diseases of the Throat and
Nose : A Manual. By Sir MORELL
MACKENZIE, M.D., Senior Physician
to the Hospital for Diseases of the Throat.
Vol. II. Diseases of the Nose and Naso-
Pharynx ; with a Section on Diseases of
the Œsophagus. Post 8vo, with 93 En-
gravings, 12s. 6d.

By the same Author.

Diphtheria :
Its Nature and Treatment, Varieties, and
Local Expressions. 8vo, 5s.

Sore Throat :
Its Nature, Varieties, and Treatment.
By PROSSER JAMES, M.D., Physician to
the Hospital for Diseases of the Throat.
Fifth Edition. Post 8vo, with Coloured
Plates and Engravings, 6s. 6d.

A Treatise on Vocal Physiology and Hygiene. By GORDON
HOLMES, M.D., Physician to the Muni-
cipal Throat and Ear Infirmary. Second
Edition, with Engravings. Crown 8vo,
6s. 6d.

A System of Dental Surgery.
By Sir JOHN TOMES, F.R.S., and C. S.
TOMES, M.A., F.R.S. Third Edition.
Crown 8vo, with 292 Engravings, 15s.

Dental Anatomy, Human and
Comparative: A Manual. By CHARLES
S. TOMES, M.A., F.R.S. Second Edition.
Crown 8vo, with 191 Engravings, 12s. 6d.

The Student's Guide to Dental
Anatomy and Surgery. By HENRY
SEWILL, M.R.C.S., L.D.S. Second
Edition. Fcap. 8vo, with 78 Engravings,
5s. 6d.

Notes on Dental Practice. By
HENRY C. QUINBY, L.D.S.R.C.S.I.
8vo, with 87 Engravings, 9s.

Mechanical Dentistry in Gold
and Vulcanite. By F. H. BALK-
WILL, L.D.S.R.C.S. 8vo, with 2 Lithographic Plates and 57 Engravings, 10s.

A Practical Treatise on Mechanical Dentistry. By JOSEPH RICH-
ARDSON, M.D., D.D.S., late Emeritus
Professor of Prosthetic Dentistry in the
Indiana Medical College. Fourth
Edition. Roy. 8vo, with 458 Engravings,
21s.

Principles and Practice of Dentistry : including Anatomy, Physiology, Pathology, Therapeutics, Dental Surgery, and Mechanism. By C. A. HARRIS, M.D., D.D.S. Edited by F. J. S. GORGAS, A.M., M.D., D.D.S., Professor in the Dental Department of Maryland University. Eleventh Edition. 8vo, with 750 Illustrations, 31s. 6d.

A Manual of Dental Mechanics.
By OAKLEY COLES, L.D.S.R.C.S. Second Edition. Crown 8vo, with 140 Engravings, 7s. 6d.

Elements of Dental Materia Medica and Therapeutics, with Pharmacopœia. By JAMES STOCKEN, L.D.S.R.C.S., Pereira Prizeman for Materia Medica, and THOMAS GADDES, L.D.S. Eng. and Edin. Third Edition. Fcap. 8vo, 7s. 6d.

Dental Medicine :
A Manual of Dental Materia Medica and Therapeutics. By F. J. S. GORGAS, A.M., M.D., D.D.S., Editor of "Harris's Principles and Practice of Dentistry," Professor in the Dental Department of Maryland University. 8vo, 14s.

Atlas of Skin Diseases.
By TILBURY FOX, M.D., F.R.C.P. With 72 Coloured Plates. Royal 4to, half morocco, £6 6s.

Diseases of the Skin :
With an Analysis of 8,000 Consecutive Cases and a Formulary. By L. D. BULKLEY, M.D., Physician for Skin Diseases at the New York Hospital. Crown 8vo, 6s. 6d.

By the same Author.

Acne : its Etiology, Pathology, and Treatment : Based upon a Study of 1,500 Cases. 8vo, with Engravings, 10s.

On Certain Rare Diseases of the Skin. BY JONATHAN HUTCHINSON, F.R.S., Senior Surgeon to the London Hospital, and to the Hospital for Diseases of the Skin. 8vo, 10s. 6d.

Diseases of the Skin :
A Practical Treatise for the Use of Students and Practitioners. By J. N. HYDE, A.M., M.D., Professor of Skin and Venereal Diseases, Rush Medical College, Chicago. 8vo, with 66 Engravings, 17s.

Parasites :
A Treatise on the Entozoa of Man and Animals, including some Account of the Ectozoa. By T. SPENCER COBBOLD, M.D., F.R.S. 8vo, with 85 Engravings, 15s.

Manual of Animal Vaccination,
preceded by Considerations on Vaccination in general. By E. WARLOMONT, M.D., Founder of the State Vaccine Institute of Belgium. Translated and edited by ARTHUR J. HARRIES, M.D. Crown 8vo, 4s. 6d.

Leprosy in British Guiana.
By JOHN D. HILLIS, F.R.C.S., M.R.I.A., Medical Superintendent of the Leper Asylum, British Guiana. Imp. 8vo, with 22 Lithographic Coloured Plates and Wood Engravings, £1 11s. 6d.

Cancer of the Breast.
By THOMAS W. NUNN, F.R.C.S., Consulting Surgeon to the Middlesex Hospital. 4to, with 21 Coloured Plates, £2 2s.

On Cancer :
Its Allies, and other Tumours; their Medical and Surgical Treatment. By F. A. PURCELL, M.D., M.C., Surgeon to the Cancer Hospital, Brompton. 8vo, with 21 Engravings, 10s. 6d.

Sarcoma and Carcinoma :
Their Pathology, Diagnosis, and Treatment. By HENRY T. BUTLIN, F.R.C.S., Assistant Surgeon to St. Bartholomew's Hospital. 8vo, with 4 Plates, 8s.

By the same Author.

Malignant Disease of the Larynx (Sarcoma and Carcinoma). 8vo, with 5 Engravings, 5s.

Also.

Operative Surgery of Malignant Disease. 8vo, 14s.

Cancerous Affections of the Skin. (Epithelioma and Rodent Ulcer.) By GEORGE THIN, M.D. Post 8vo, with 8 Engravings, 5s.

By the same Author.

Pathology and Treatment of Ringworm. 8vo, with 21 Engravings, 5s.

Cancer of the Mouth, Tongue, and Alimentary Tract : their Pathology, Symptoms, Diagnosis, and Treatment. By FREDERIC B. JESSETT, F.R.C.S., Surgeon to the Cancer Hospital, Brompton. 8vo, 10s.

Clinical Notes on Cancer,
Its Etiology and Treatment ; with special reference to the Heredity-Fallacy, and to the Neurotic Origin of most Cases of Alveolar Carcinoma. By HERBERT L. SNOW, M.D. Lond., Surgeon to the Cancer Hospital, Brompton. Crown 8vo, 3s. 6d.

Lectures on the Surgical Disorders of the Urinary Organs. By REGINALD HARRISON, F.R.C.S., Surgeon to the Liverpool Royal Infirmary. Third Edition, with 117 Engravings, 8vo, 12s. 6d.

Hydrocele :
Its several Varieties and their Treatment. By SAMUEL OSBORN, late Surgical Registrar to St. Thomas's Hospital. Fcap. 8vo, with Engravings, 3s.

By the same Author.

Diseases of the Testis.
Fcap. 8vo, with Engravings, 3s. 6d.

Diseases of the Urinary Organs.
Clinical Lectures. By Sir HENRY THOMPSON, F.R.C.S., Emeritus Professor of Clinical Surgery in University College. Seventh (Students') Edition. 8vo, with 84 Engravings, 2s. 6d.

By the same Author.

Diseases of the Prostate:
Their Pathology and Treatment. Sixth Edition. 8vo, with 39 Engravings, 6s.

Also.

Surgery of the Urinary Organs.
Some Important Points connected therewith. Lectures delivered in the R.C.S. 8vo, with 44 Engravings. Students' Edition, 2s. 6d.

Also.

Practical Lithotomy and Lithotrity; or, An Inquiry into the Best Modes of Removing Stone from the Bladder. Third Edition. 8vo, with 87 Engravings, 10s.

Also.

The Preventive Treatment of Calculous Disease, and the Use of Solvent Remedies. Second Edition. Fcap. 8vo, 2s. 6d.

Also.

Tumours of the Bladder:
Their Nature, Symptoms, and Surgical Treatment. 8vo, with numerous Illustrations, 5s.

Also.

Stricture of the Urethra, and Urinary Fistulæ: their Pathology and Treatment. Fourth Edition. With 74 Engravings. 8vo, 6s.

Also.

The Suprapubic Operation of Opening the Bladder for the Stone and for Tumours. 8vo, with 14 Engravings, 3s. 6d.

The Surgery of the Rectum.
By HENRY SMITH, Professor of Surgery in King's College, Surgeon to the Hospital. Fifth Edition. 8vo, 6s.

Modern Treatment of Stone in the Bladder by Litholopaxy. By P. J. FREYER, M.A., M.D., M.Ch., Bengal Medical Service. 8vo, with Engravings, 5s.

Diseases of the Testis, Spermatic Cord, and Scrotum. By THOMAS B. CURLING, F.R.S., Consulting Surgeon to the London Hospital. Fourth Edition. 8vo, with Engravings, 16s.

Diseases of the Rectum and Anus. By W. HARRISON CRIPPS, F.R.C.S., Assistant Surgeon to St. Bartholomew's Hospital, &c. 8vo, with 13 Lithographic Plates and numerous Wood Engravings, 12s. 6d.

Urinary and Renal Derangements and Calculous Disorders.
By LIONEL S. BEALE, F.R.C.P., F.R.S., Physician to King's College Hospital. 8vo, 5s.

Fistula, Hæmorrhoids, Painful Ulcer, Stricture, Prolapsus, and other Diseases of the Rectum:
Their Diagnosis and Treatment. By WILLIAM ALLINGHAM, Surgeon to St. Mark's Hospital for Fistula. Fourth Edition. 8vo, with Engravings, 10s. 6d.

Pathology of the Urine.
Including a Complete Guide to its Analysis. By J. L. W. THUDICHUM, M.D., F.R.C.P. Second Edition, rewritten and enlarged. 8vo, with Engravings, 15s.

Student's Primer on the Urine.
By J. TRAVIS WHITTAKER, M.D., Clinical Demonstrator at the Royal Infirmary, Glasgow. With 16 Plates etched on Copper. Post 8vo, 4s. 6d.

Syphilis and Pseudo-Syphilis.
By ALFRED COOPER, F.R.C.S., Surgeon to the Lock Hospital, to St. Mark's and the West London Hospitals. 8vo, 10s. 6d.

Diagnosis and Treatment of Syphilis. By TOM ROBINSON, M.D., Physician to St. John's Hospital for Diseases of the Skin. Crown 8vo, 3s. 6d.

By the same Author.

Eczema: its Etiology, Pathology, and Treatment. Crown 8vo, 3s. 6d.

Coulson on Diseases of the Bladder and Prostate Gland.
Sixth Edition. By WALTER J. COULSON, Surgeon to the Lock Hospital and to St. Peter's Hospital for Stone. 8vo, 16s.

The Medical Adviser in Life Assurance. By Sir E.H. SIEVEKING, M.D., F.R.C.P. Second Edition. Crown 8vo, 6s.

A Medical Vocabulary:
An Explanation of all Terms and Phrases used in the various Departments of Medical Science and Practice, their Derivation, Meaning, Application, and Pronunciation. By R. G. MAYNE, M.D., LL.D. Fifth Edition. Fcap. 8vo, 10s. 6d.

A Dictionary of Medical Science:
Containing a concise Explanation of the various Subjects and Terms of Medicine, &c. By ROBLEY DUNGLISON, M.D., LL.D. Royal 8vo, 28s.

Medical Education
And Practice in all parts of the World. By H. J. HARDWICKE, M.D., M.R.C.P. 8vo, 10s.

—◆—

The following CATALOGUES issued by J. & A. CHURCHILL will be forwarded post free on application :—

A. *J. & A. Churchill's General List of about 650 works on Anatomy, Physiology, Hygiene, Midwifery, Materia Medica, Medicine, Surgery, Chemistry, Botany, &c., &c., with a complete Index to their Subjects, for easy reference. N.B.—This List includes* B, C, & D.

B. *Selection from J. & A. Churchill's General List, comprising all recent Works published by them on the Art and Science of Medicine.*

C. *J. & A. Churchill's Catalogue of Text Books specially arranged for Students.*

D. *A selected and descriptive List of J. & A. Churchill's Works on Chemistry, Materia Medica, Pharmacy, Botany, Photography, Zoology, the Microscope, and other branches of Science.*

E. *The Half-yearly List of New Works and New Editions published by J. & A. Churchill during the previous six months, together with particulars of the Periodicals issued from their House.*

[Sent in January and July of each year to every Medical Practitioner in the United Kingdom whose name and address can be ascertained. A large number are also sent to the United States of America, Continental Europe, India, and the Colonies.]

AMERICA.—*J. & A. Churchill being in constant communication with various publishing houses in Boston, New York, and Philadelphia, are able, notwithstanding the absence of international copyright, to conduct negotiations favourable to English Authors.*

LONDON: 11, NEW BURLINGTON STREET.

www.ingramcontent.com/pod-product-compliance
Lightning Source LLC
Chambersburg PA
CBHW021709210326
41599CB00013B/1582